Biochemistry of the Phagocytic Process

Biochemistry of the Phagocytic Process

Localization and the Role of Myeloperoxidase and
the Mechanism of the Halogenation Reaction

Editor

JULIUS SCHULTZ

Papanicolaou Cancer Research Institute, and
University of Miami, School of Medecine, Miami, Fla.

1970

NORTH-HOLLAND PUBLISHING COMPANY - AMSTERDAM · LONDON

Library of Congress Catalog Card Number: 73-108277

ISBN 7204 4059 9

Publishers:

NORTH-HOLLAND PUBLISHING COMPANY – AMSTERDAM
NORTH-HOLLAND PUBLISHING COMPANY, Ltd. – LONDON

PRINTED IN THE NETHERLANDS

PREFACE

In January, 1969, a Symposium was held in two parts; that organized by the Department of Biochemistry of the University of Miami Medical School and that organized by the Papanicolaou Cancer Research Institute of Miami. The present Symposium includes only that of the latter inasmuch as commitments were not obtained from the speakers for the former. In subsequent years both Symposia will be published under a single cover.

The chemistry of peroxidase in the past has been mostly concerned with the mechanism of action and the properties of heme enzymes. There was an unusual opportunity at this Symposium to bring together investigators whose researches had to do with the particular properties of peroxidases related to the halogenation reaction, at a time when a great deal of attention was being centered on the work of Klebanoff on the bacteriocidal effects of this reaction. More striking was the recent discovery of two diseases which for the first time made it possible to assign a biological function to a peroxidase, specifically the peroxidase of the leucocyte, that is myeloperoxidase. Thus in the first series of papers, the property of the halogenation reaction and two well-known peroxidases are described. New data and information about the chemical properties of myeloperoxidase and its functional origin, the lysosomes, are reported. The variety of lysosomes in view of the recog-

5

nized heterogeneity of this cellular particulate, and finally, the description of an amyeloperoxidase patient is given, with evidence of a lack of myeloperoxidase in the patient's white cells.

What becomes clearly evident is that the phagocytic process which begins with DPNH oxidase results in the formation of hydrogen peroxide which is required for the bacteriocidal action of myeloperoxidase coming from the ruptured granules. In those cases of children's granulomatous disease where the lysosome fails to rupture, killing can not take place, in which case it has been demonstrated that there is a lack of DPNH oxidase enzyme. Most exciting in this connection is a report by Klebanoff that peroxide generating organisms are phagocytized and killed by white cells of patients with childrens' granulomatous disease. Peroxidase is released from the granuals for killing only in the case of live organisms whereas the dead organism does not cause rupture of the granules, although in normal cells the dead organism can result in rupturing the granule with release of the enzyme.

The series of papers presented here are not only novel in themselves, but represent a 'first' in the total story being presented in the phagocytic process only now being unfolded in chemical terms.

TABLE OF CONTENTS

EVIDENCE FOR A MEMBRANE-LINKED Ca^{++} CARRIER IN RAT LIVER AND KIDNEY MITOCHONDRIA

Albert L. LEHNINGER and Ernesto CARAFOLI

Department of Physiological Chemistry
The Johns Hopkins University
School of Medicine
725 North Wolfe Street
Baltimore, Maryland 21205, USA

This paper is concerned with the dynamics of Ca^{++} movements coupled to the energy-conserving mechanisms in mitochondria. It also summarizes evidence for the occurrence of a Ca^{++} carrier or "permease" in the inner membrane of liver and kidney mitochondria.

It is now well established that Ca^{++} uptake, as well as uptake of Mn^{++} and Sr^{++}, is coupled to electron transport [1]. When phosphate is the external anion both Ca^{++} and phosphate are accumulated in a fixed stoichiometric ratio in relation to electron transport; the Ca^{++} : $\sim$ ratio is then about 2.0. When phosphate or other permeant anions are absent from the medium, Ca^{++} accumulation is accompanied by ejection of H$^+$ ions (H$^+$: $\sim$ $\leqslant$ 2.0) and alkalinization of the mitochondrial membrane (OH$^-$: $\sim$ $\leqslant$ 2.0) [2–5]. However, in the absence of permeant anions, the Ca^{++} : $\sim$ ratio is not fixed. When the pH is 7.4 and KCl concentration is 80 mM, the

Ca^{++} : ~ ratio is about 2.0, but when pH or KCl concentration or both are increased, the Ca^{++} : ~ ratio also increases, and may reach values as high as 20 [6,7]. Such "superstoichiometry" of Ca^{++} uptake has been attributed to an abnormal ion distribution on, near, or across the membrane, inducing a state in which anionic sites on the membrane can bind much more Ca^{++} than usual. In this situation electron transport is also severely inhibited [6–8], a condition known as State 6 [8].

There is another set of circumstances in which Ca^{++} uptake is not stoichiometric with electron transport. At low pH and KCl concentrations, the uptake of Ca^{++} and ejection of H^+ during respiration may show reversals or oscillations without corresponding changes in the rate of electron transport [9].

The occurrence of superstoichiometry and of $Ca^{++} - H^+$ oscillations which are non-stoichiometric with electron transport strongly suggest that entry of Ca^{++} is not always strictly and stoichiometrically dependent on electron transport. We have therefore postulated that Ca^{++} (and also Mn^{++} and Sr^{++}) enter and leave mitochondria through the action of a specific carrier or permease for Ca^{++}, Mn^{++}, and Sr^{++} [9–11], analogous to the carriers for ATP, succinate, and other anions. By analogy with other passive carriers, the postulated Ca^{++} carrier would be expected to show specificity for the cation transported, saturability, independence from electron transport and oxidative phosphorylation, and the capacity to function in either direction. It would not be expected to transport Ca^{++} against a gradient unless a source of energy is available.

More recently additional evidence supporting the existence of a specific Ca^{++} carrier in the inner mitochondrial membrane has arisen from a more detailed analysis of respiration-independent binding of Ca^{++} by rat liver mitochondria [10]. Azzone and his colleagues [12] had earlier shown that rat liver mitochondria bind up to 50 nmoles Ca^{++} per mg protein when respiration is inhibited. However, they found that the affinity of the binding sites for Ca^{++} was quite low, since a concentration of at least 200 μM Ca^{++} was required to half-saturate them. Such low-affinity binding is probably not a step in the energy-dependent uptake of Ca^{++}, which has long been known to proceed maximally at much lower concentrations of Ca^{++}. In more detailed studies by Reynafarje and Lehninger [10,11], it was shown that rat liver mitochondria possess two different sets of binding sites for Ca^{++} when respiration was inhibited by antimycin A + rotenone. One set, relatively large in number (30–40 nmoles per mg protein) but low in affinity for Ca^{++}, corresponding to the binding sites for Ca^{++}, was very small in number, at most about 5 nmoles per mg protein, but very high in affinity ($K = 0.1 \mu$M). The occurrence of the two sets of binding sites was revealed by Scatchard plots of Ca^{++} binding by intact rat liver mitochondria at low Ca^{++} concentrations. The high-affinity binding sites could be filled by Ca^{++}, Mn^{++}, and Sr^{++}, but not by Mg^{++} or K^{+}. Neither Mg^{++}, K^{+}, nor Na^{+} competed with Ca^{++}. Binding of Ca^{++} to either the high-affinity or low-affinity binding sites was unaccompanied by H^{+} release [10], in sharp contrast to respiration-dependent Ca^{++} uptake, which is accompanied by mole-for-mole H^{+} ejection (cf. [1]).

The high-affinity respiration-independent binding of Ca^{++} by rat liver mitochondria was found to be inhibited by uncoupling agents such as 2,4-dinitrophenol, dicumarol, and FCCP, but was not influenced by oligomycin, aurovertin, valinomycin, or gramicidin + NaCl. High-affinity Ca^{++} binding is also inhibited by the local anesthetic butacaine, which has been shown to displace Ca^{++} from phospholipid micelles [13]. Mela has also observed effects of butacaine on accumulation of Ca^{++} by mitochondria [14]. In the light of earlier work carried out in our laboratory by Neubert on the characteristic effects of rare earth cations on mitochondrial lipids and ATPase activity [15], we examined a number of rare earth and other metal ions for their capacity to prevent high-affinity binding of Ca^{++}. Among these ions, La^{3+} gave the most striking inhibition of directly measured Ca^{++} binding, being effective at $1-10$ μM. This finding concurs with the independent spectroscopic observation of Mela that La^{+++} interferes in the interaction of Ca^{++} with mitochondria [14].

The characteristic inhibition of high-affinity Ca^{++} binding by uncoupling agents was accompanied by very fast discharge of some of the endogenous Ca^{++} of mitochondria [11]. It is of some interest that the labile portion of the pool of endogenous Ca^{++} of mitochondria ($5-7$ nmoles per mg protein) is rapidly discharged by uncoupling agents, but not by respiratory inhibitors. It was tentatively concluded that binding of Ca^{++} to the high-affinity sites may require the pre-existence of a high-energy state, either a chemical intermediate $X \sim I$, postulated by the chemical coupling hypothesis, or a pre-existing H^+ ion gradient, such as postulated by the chemiosmotic hypothesis.

High-affinity Ca^{++} binding is labile to ageing of mitochondria at room temperature and to heating to $60°$; low-affinity Ca^{++} binding is not affected by these treatments. High-affinity Ca^{++} binding is thus not merely a non-specific effect given by a phospholipid or protein. A second major conclusion is that high-affinity Ca^{++} binding may be selectively inactivated without inactivation of the capacity for oxidative phosphorylation or to accumulate Ca^{++}. Water lysis of mitochondria to yield ghosts causes complete loss of high-affinity Ca^{++} binding, but capacity for oxidative phosphorylation ($P:O \leqslant 2.0$) is retained. Such ghosts also retain the capacity for respiration-linked Ca^{++} uptake, but they require the presence of phosphate and ADP [16]. Treatment of mitochondria with the detergent Lubrol also causes loss of high affinity binding. On the other hand, mitochondrial membrane particles made by sonic methods or with digitonin retain full capacity for high-affinity Ca^{++} binding. The capacity for high affinity Ca^{++} binding is apparently localized in the inner membrane [10,11].

To account for these observations Reynafarje and Lehninger [10] postulated that high-affinity binding of Ca^{++} is a reflection of the presence of a specific membrane carrier or permease capable of passively transferring Ca^{++}, Sr^{++}, or Mn^{++} across the inner mitochondrial membrane, analogous to the several anion permeases which have been identified in mitochondria, such as those for ATP, succinate, and isocitrate [17]. The Ca^{++} carrier was postulated to have a very high affinity for Ca^{++}, comparable to and even exceeding the affinity of most enzymes for their substrates, thus accounting for the extremely high affinity of the respiratory chain for Ca^{++}. However, when the Ca^{++} carrier is damaged selectively, as happens on treatment with Lubrol or hypotonic conditions, then Ca^{++} may still be accumulated by mitochondria, but it must then enter by a non-facilitated concentration-dependent diffusion through the membrane. It was further suggested that the carrier can operate independently of electron transport, to account for the fact that Ca^{++} can oscillate back and forth across the membrane without change in the rate of electron transport. It was also suggested that the net transfer of Ca^{++} into and out of mitochondria takes place with compensatory inverse movement of other cations or simultaneous movement of counter-anions with Ca^{++}, in order to preserve the balance of electrical charges across the membrane.

Two experimental approaches are currently being taken to investigate the Ca^{++} carrier further. In one, we are attempting to isolate a solubilized Ca^{++} binding component from rat liver mitochondria which possesses the same extremely high affinity for Ca^{++} shown by intact mitochondria. This approach is not yet successful; an obvious experimental difficulty is the relative instability of the high-affinity binding sites for Ca^{++}. However, the other approach we are taking, a comparative study of Ca^{++} binding, is turning up some fruitful leads, and will be discussed in more detail.

Chappell and others [17] have shown that the anion permeases specific for succinate and isocitrate, which are present in rat liver and kidney mitochondria, are lacking in blowfly muscle mitochondria, which are unable to oxidize externally added tricarboxylic acid cycle intermediates so long as their membranes are intact. However, following treatments that increase the permeability of the membrane to these anions, blowfly mitochondria readily oxidize the cycle intermediates at a high rate. It therefore appears likely that the various mitochondrial permeases may be genetically determined and that their distribution in mitochondria of different cell types may be a reflection of the genotype. Presumably the intensely respiring blowfly mitochondria contain no permeases so that they can maintain extremely high internal concentrations of Krebs cycle intermediates, particularly of isocitrate, the usual rate-limiting factor, for maximal rates of respiration.

We therefore set about to survey Ca^{++} accumulation and high-affinity binding capacity in a variety of different mitochondria, chosen to represent widely different cell types of different metabolic characteristics, in the hope that some mitochondrial types might show the presence of the imputed Ca^{++} carrier and others might not, in order to add comparative biological evidence to the evidence already at hand regarding rat liver mitochondria. Actually, a survey of Ca^{++} transport characteristics in various mitochondria badly needed doing in any case, if for no other reason than to check whether stoichiometric Ca^{++} accumulation is linked to electron transport as universally as is ADP phosphorylation.

Our survey of various mitochondria [18,19] has in fact yielded a number of hitherto unsuspected differences, and the remainder of this paper will describe our results on mitochondria from rat liver, kidney, heart, *Neurospora crassa*, yeast, and blowfly muscle. We are indebted to Dr. John Greenawalt for preparation of *Neurospora* mitochondria, to Dr. James Mattoon for preparation of wild-type and mutant mitochondria of yeast, and to Dr. Bertram Sacktor for preparations of blowfly mitochondria.

We soon found that these mitochondria differ widely in endogenous Ca^{++} content when isolated in an EDTA-free 0.3 M sucrose medium (table 1). Yeast, which does not usually require Ca^{++} for growth, contains relatively little endogenous Ca^{++} in its mitochondria, whereas blowfly mitochondria contain very large amounts. It must be remembered that blowfly flight muscle is asynchronous and contains no sarcoplasmic reticulum for reversible segregation of Ca^{++}.

The second major observation is that not all mitochondria are

Table 1
Ca^{++} content of freshly isolated mitochondria
(no EDTA in isolation medium

Mitochondria	nmoles Ca^{++} per mg protein
Rat liver	10–14
Rat kidney	35–55
Rat heart	30–40
Neurospora	20
Yeast	8–9
Blowfly muscle	40–60

capable of accumulating Ca^{++} in a respiration-dependent process. Data in fig. 1 show that although rat liver, kidney, and heart mitochondria, as well as *Neurospora*, do accumulate Ca^{++} in the presence of phosphate in a dinitrophenol-sensitive process, yeast mitochondria do not. Yeast mitochondria can take up large amounts of added Ca^{++}, but this process is not inhibited by either DNP or respiratory inhibitors. We have also grown yeast on a Ca^{++}-free medium; mitochondria isolated from these cells failed to show Ca^{++} uptake, in the

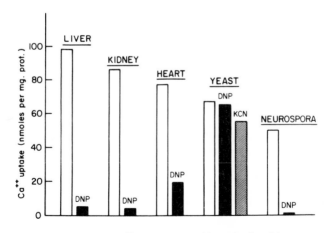

Fig. 1. Effect of DNP on Ca^{++} uptake by respiring mitochondria.

absence or presence of phosphate. Clearly, yeast mitochondria contain the capacity to bind larger amounts of Ca^{++} in a low-affinity,
energy-independent fashion, than do other types of mitochondria.
In order to establish a possible molecular basis for this observation,
we are carrying out studies of mitochondrial lipids and protein
composition in yeast and other mitochondria.

A third major observation is that Ca^{++} does not necessarily stim-
ulate State 4 respiration of those mitochondria which can accumulate Ca^{++} in a respiration-dependent process. Although blowfly
mitochondria can accumulate Ca^{++} in a dinitrophenol-sensitive process, Ca^{++} addition has absolutely no effect on the State 4 respiratory rates (fig. 2). Yet under these conditions Ca^{++} is accumulated
and its accumulation is blocked by dinitrophenol. The $Ca^{++} : \sim$
accumulation ratio, however, is relatively low, about 0.6, in con-

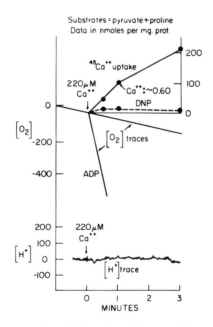

Fig. 2. DNP-sensitive Ca^{++} uptake without respiratory stimulation
in blowfly muscle mitochondria.

trast to normal values in rat liver mitochondria of about 2.0. These observations are reminiscent of the State 4 accumulation of Ca^{++} we described earlier in rat liver mitochondria, which proceeded with a Ca^{++} : $\sim$ ratio of about 0.2 [20]. We will return to the significance of the blowfly observations later.

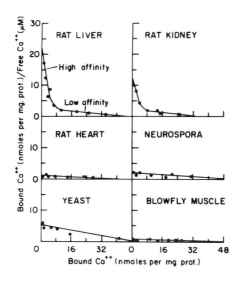

Fig. 3. Comparison of high-affinity and low-affinity Ca^{++} binding
(Scatchard plots).

A fourth major observation made in this comparative study is that all mitochondria show respiration-independent low-affinity binding of Ca^{++} (fig. 3). While the number and affinity of these sites vary somewhat, in all cases the Scatchard plots for low-affinity binding are rectilinear, showing that there is no cooperativity of interaction among the low affinity sites. What is remarkable about the low-affinity sites is that they are differently sensitive to dinitrophenol. Those mitochondria showing the presence of high-affinity Ca^{++} binding, namely liver and kidney, also show DNP-sensitive low-affinity Ca^{++} binding. On the other hand, yeast, blowfly, and rat heart mitochondria show no DNP-sensitive low-affinity Ca^{++} binding.

The fifth observation made in this comparative study, which also is germane to the question of the presence of Ca^{++} permease, is that not all mitochondria show high-affinity Ca^{++} binding. Actually only rat liver and kidney mitochondria, of all those examined to date, show high-affinity binding as reflected in biphasic Scatchard plots. All other mitochondria show monophasic, completely rectilinear Scatchard plots. A basic premise assumed at the outset of this comparative study is therefore fulfilled; there appears to be a day-and-night difference among mitochondria from various cells as to their capacity for high affinity Ca^{++} binding and thus, possibly in their possession of a high-affinity permease or carrier for Ca^{++}. Our findings do not exclude the possibility that the other types of mitochondria may contain permeases which have very low affinity for Ca^{++}; such permeases would not be detected by Scatchard plots.

In confirmation of this hypothesis we have found that maximal rates of Ca^{++} uptake and maximal rates of stimulated respiration in rat liver and kidney are produced by much lower concentrations of Ca^{++} than is the case for rat heart mitochondria, which show no high-affinity Ca^{++} binding, an observation that is fully consistent with the view that entrance of Ca^{++} in liver mitochondria is brought about by a carrier. It appears to be a reasonable hypothesis that all mitochondria intrinsically possess the capacity to drive the accumulation of Ca^{++}, providing some mechanism exists to allow Ca^{++} to penetrate the membrane. In the case of rat liver and kidney mitochondria, the imputed Ca^{++} carrier provides access of Ca^{++} to the interior. In the case of other types of mitochondria, such as heart, a slower, concentration-dependent diffusion of Ca^{++} may be the pathway of entry. Yeast mitochondria can evidently bind large amounts of Ca^{++}, but very likely this is surface binding.

The question may now be raised: What is the immediate driving force of Ca^{++} accumulation? Is it a high energy chemical inter-

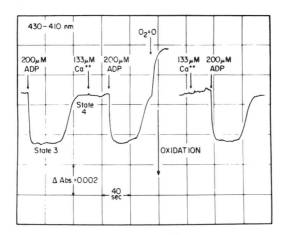

Fig. 4. Response of cytochrome b to ADP and to Ca^{++} in yeast mitochondria.

mediate, a derivative of one of the electron transport carriers, is it a H$^+$ ion gradient across the membrane, or is it a specific conformational change in the membrane? We do not have the answer to these questions. However, we have made some observations relevant to this question. It will be recalled that Chance and his colleagues have shown that one of the most immediate, if not the most immediate effect of Ca^{++} on mitochondria, is its capacity to cause a shift in the steady state of cytochrome b to a more oxidized condition [21]. Dr. Balcavage in our laboratory has carried out some double-beam measurements of the effect of Ca^{++} on the state of cytochrome b in yeast mitochondria (fig. 4), which shows that although ADP addition yields the usual shift of cytochrome b to a more oxidized state, the addition of Ca^{++} produced no such effect. Either Ca^{++} does not reach cytochrome b or the cytochrome b of yeast mitochondria is unreactive with Ca^{++}. When ADP is added to blowfly mitochondria, the usual shift toward a more oxidized state of cytochrome b occurs. However, addition of Ca^{++} produced no change whatsoever in the oxidation-reduction state of cytochrome b, even though the mitochondria were accumulating

Ca^{++} in a DNP-dependent process. These experiments raise some question as to whether the Ca^{++}–cytochrome b interaction is indeed the "driving" step in energy-linked Ca^{++} accumulation.

We have also found that some mitochondria, such as those from blowfly muscle, require phosphate for accumulation of Ca^{++} and will in fact fail to accumulate Ca^{++} if the phosphate carrier is blocked by mersalyl. This finding recalls the observations of Hanson and Miller in maize mitochondria [22], as well as those of Bonner and Pressman [23], which also indicate a phosphate requirement. We may yet have to reconsider the old question of what is transported first, cations or anions.

We may now summarize the major points of evidence supporting the view that the inner membrane of some mitochondria possess a carrier or permease specific for Ca^{++} and certain other divalent metal ions:

1. Respiration-dependent Ca^{++} uptake is saturable with Ca^{++}. This fact clearly differentiates this process from simple physical diffusion, which would be expected to be completely concentration-dependent. Saturability infers that the binding site of the carrier reversibly binds Ca^{++} in a simple equilibrium process. This criterion has already been used to identify carriers for glucose, galactose and other substrates in other types of cells.

2. The affinity of the high affinity Ca^{++} binding sites is comparable to and indeed greater than the affinity of most carriers or enzymes for their substrate.

3. The permease or binding site has specificity. It binds Ca^{++}, Sr^{++} and Mn^{++}, which compete with each other, but does not bind Mg^{++} or Ba^{++}. It also binds La^{+++}, but it has not been clearly established whether La^{+++} binding is competitive.

4. The Ca^{++} carrier can function independently of electron transport, as appears to occur in the phenomenon of superstoichiometry and the inverse Ca^{++}–H^+ oscillations.

5. The Ca^{++} carrier may be selectively inactivated by ageing or osmotic shock without inactivation of oxidative phosphorylation.

6. The Ca^{++} carrier is not required for Ca^{++} uptake. Presumably Ca^{++} can cross the mitochondrial membrane by simple diffusion as well as via the carrier, but the rate of the non-facilitated movement is much lower than the facilitated movement.

7. The Ca^{++} carrier is present in only certain mitochondria, presumably because it is genetically determined. This observation is consistent with observations on other permeases. For example, a glucose carrier is known to be present in human erythrocytes, but is absent in bovine erythrocytes.

Although these findings satisfy all major criteria that have been recognized for the occurrence of membrane-linked permeases or carriers, it would of course be most desirable to have more direct evidence as to the molecular nature of the Ca^{++} carrier. Recent work has shown that a specific Ca^{++}-binding protein from the small intestine [24], is able to bind Ca^{++} with high affinity ($K_{Ca^{++}} = 10^5$ M). However, the affinity of rat liver mitochondria for Ca^{++} is at least 10-fold greater. The isolation and proof of identity of a permease is still a very difficult business at best, as recent work on the sulfate, amino acid and glucose permeases of bacteria has shown. Nevertheless we are continuing our efforts to isolate and purify the imputed carriers for both ATP and Ca^{++}.

References

[1] A.L.Lehninger, E.Carafoli and C.S.Rossi, Energy-linked ion accumulation in mitochondrial systems, Advan. Enzymol. 29 (1967) 259.
[2] C.S.Rossi, J.Bielawski and A.L.Lehninger, Separation of H$^+$ and OH$^-$ in the extramitochondrial phases during Ca^{++} activated electron transport, J. Biol. Chem. 241 (1966) 1919–1921.
[3] A.R.L.Gear, C.S.Rossi, B.Reynafarje and A.L.Lehninger, Acid-base exchanges in mitochondria and suspending medium during respiration-linked accumulation of bivalent cations, J. Biol. Chem. 242 (1967) 3403–3413.
[4] B.Chance and L.Mela, Hydrogen ion concentration changes in mitochondrial membranes, J. Biol. Chem. 241 (1966) 4588–4599.
[5] S.Addanki, C.F.Dallas and J.F.Sotos, Determination of intramitochondrial pH and intramitochondrial–extramitochondrial pH gradient of isolated heart mitochondria by the use of 5,5-dimethyl-2,4-oxazolidinedione, J. Biol. Chem. 243 (1968) 2337–2348.

[6] E.Carafoli, R.L.Gamble, C.S.Rossi and A.L.Lehninger, Super-stoichiometric ratios between ion movements and electron transport in rat liver mitochondria, J. Biol. Chem. 242 (1967) 1199–1204.

[7] C.S.Rossi and G.F.Azzone, $H^+:O$ ratio during Ca^{++} uptake in rat liver mitochondria, Biochim. Biophys. Acta 110 (1965) 434–436.

[8] B.Chance and B.Schoener, High and low energy states of cytochromes in reactions with cations, J. Biol. Chem. 241 (1966) 4577–4587.

[9] E.Carafoli, R.L.Gamble and A.L.Lehninger, Rebounds and oscillations in respiration-linked movements of Ca^{++} and H^+ in rat liver mitochondria, J. Biol. Chem. 241 (1966) 2644–2652.

[10] B.Reynafarje and A.L.Lehninger, High-affinity and low-affinity binding of Ca^{++} by rat liver mitochondria, J. Biol. Chem. 244 (1969) 584–593.

[11] A.L.Lehninger, Acid-base changes in mitochondria and medium during energy-dependent and energy-independent binding of Ca^{++}, Ann. N.Y. Acad. Sci. (1969), in press.

[12] C.Rossi, A.Azzi and G.F.Azzone, Ion transport in liver mitochondria. I. Metabolism-independent Ca^{++} binding and H^+ release, J. Biol. Chem. 242 (1967) 951–957.

[13] M.Feinstein, Reactional local anesthetics with phospholipids, J. Gen. Physiol. 48 (1964) 357–374.

[14] L.Mela, Interactions of La^{3+} and local anesthetic drugs with mitochondrial Ca^{++} and Mn^{++} uptake, Arch. Biochem. Biophys. 123 (1968) 286–293.

[15] D.Neubert, Mitochondrial formation and hydrolysis of ATP in the presence of some rare-earth ions, Biochim. Biophys. Acta 69 (1963) 399–402.

[16] A.Caplan and J.W.Greenawalt, Biochemical and ultrastructural properties of osmotically lysed rat-liver mitochondria, J. Cell Biol. 31 (1966) 455–472.

[17] J.B.Chappell, Systems used for the transport of substrates into mitochondria, Brit. Med. Bull. 24 (1968) 150–157.

[18] A.L.Lehninger and E.Carafoli, The reaction of Ca^{++} with the mitochondrial membrane: A comparative study of mitochondria from different sources, Fed. Proc. 28 (1969) 664.

[19] E.Carafoli, J.Mattoon, W.Balcavage and A.L.Lehninger, A comparative study of the interaction of Ca^{++} with mitochondria from different cell types, in preparation.

[20] E.Carafoli, C.S.Rossi and A.L.Lehninger, Energy coupling in mitochondria during resting or State 4 respiration, Biochem. Biophys. Res. Commun. 19 (1965) 609–614.

[21] B.Chance, The energy linked reaction of calcium with mitochondria, J. Biol. Chem. 240 (1965) 2729–2748.

[22] J.B.Hanson and R.J.Miller, Evidence for active phosphate transport in maize mitochondria, Proc. Natl. Acad. Sci. U.S. 58 (1967) 727–734.

[23] W.D.Bonner and B.C.Pressman, Plant Physiol. 40 (1965) lv–lvi.

[24] R.H.Wasserman, R.A.Corradino and A.N.Taylor, Vitamin D-dependent calcium-binding protein. I. Purification and some properties. II. Response to some physiological and nutritional variables, J. Biol. Chem. 243 (1968) 3978–3986, 3987–3993.

THE OXIDATION OF CYTOCHROME c CATALYZED BY CYTOCHROME OXIDASE AND HORSERADISH AND CYTOCHROME c PEROXIDASES *

Takashi YONETANI **, Toshio ASAKURA,
Henry R. DROTT ‡ and C. P. LEE ‡‡

*The Johnson Research Foundation, University of Pennsylvania,
Philadelphia, Pa. 19104, USA*

1. Introduction

The function of cytochrome c as one of the essential electron carriers that link the dehydrogenase system to the oxidase system was established in the 1930's by Keilin [1]. Keilin [2], Theorell [3], and their associates did pioneer work on enzymic and physical chemical properties of cytochrome c. Today the primary structures of cytochrome c from various sources have been established by

* Supported by research grants from the National Science Foundation (GB 6974) and the United States Public Health Service (GM 12202).
** Recipient of Career Development Award 1-K3-GM35,331 from the United States Public Health Service.
‡ Recipient of Postdoctoral Fellowship 1-F2-HE39,533 from the United States Public Health Service.
‡‡ Recipient of Career Development Award 1-K4-38,822 from the United States Public Health Service.

Smith [4], Margoliash [5], and their associates. The three-dimensional structure of the cytochrome has been worked out by Dickerson et al. [6] by X-ray crystallography. In addition, I would like to mention that Sano of Kyoto University has just succeeded in the complete chemical reconstitution of cytochrome c from individual amino acids and heme [7]. Sano showed that the chemically synthesized cytochrome c is enzymically active. This indicates that conformational information is already stored in the primary sequence of amino acids in this case. Recent studies on the reactivity of cytochrome c in enzyme and membrane systems in the Johnson Foundation was reviewed in detail by Chance in the Second Keilin Memorial Lecture [8].

Today we would like to discuss our more recent studies on the spin labeled cytochrome c [9] and the mechanism of cytochrome c oxidation catalyzed by cytochrome oxidase and cytochrome c peroxidase.

2. Results and discussion

A stable free radical, N-oxyl-tetramethyl-pyrrolidinyl bromoacetamide, has been chemically incorporated into cytochrome c [9]. This spin-labeled cytochrome c retains optical and enzymic properties similar to the native cytochrome c. As shown in fig. 1, the spin-label cytochrome c exhibits EPR spectra of the weakly immobilized spin label, indicating that the alkylation occurs on the surface of the molecule. The EPR spectrum of spin-labeled cytochrome c is relatively insensitive to oxidation-reduction of the cytochrome, whereas it is highly influenced by interactions with cytochrome c-specific enzymes such as cytochrome oxidase and cytochrome c peroxidase and mitochondrial membrane fragments. For example, when the spin-labeled cytochrome c is bound to beef heart EDTA particles of Lee and Ernster [10], EPR signals are considerably broadened, indicating that the conformational flexibility of the cytochrome is more restricted upon binding to the particles. The

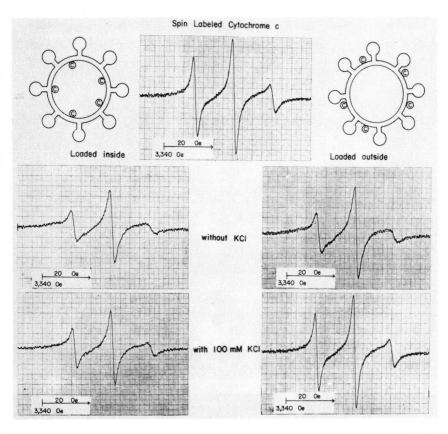

Fig. 1. EPR spectra of spin-labeled cytochrome *c* bound and unbound to mitochondrial membrane fragments. Temperature, 20°; microwave frequency, 9.0 GHz.

spin-labeled cytochrome *c* which is tightly bound to the inner surface of the particles, or the inside loaded cytochrome *c*, exhibits oligomycin-induced control of electron flow during oxidative phosphorylation. We are currently investigating the conformation change of the membrane fragments associated with the energized state by the use of the EPR technique. The inside-loaded cytochrome *c* is insensitive to the KCl treatment. However, the outside-loaded spin-labeled cytochrome *c* can be readily released from the EDTA particles upon the addition of KCl.

The oxidation of ferrocytochrome c is catalyzed by two mito-chondrial enzymes, namely, cytochrome oxidase and cytochrome c peroxidase. These enzymes were discovered in the late 1930's by Keilin and Hartree [11] and Altschul, Abrams, and Hogness [12], respectively. Chance [13] showed that a variety of peroxidases from plant and animal sources can also oxidize cytochrome c at much slower rates.

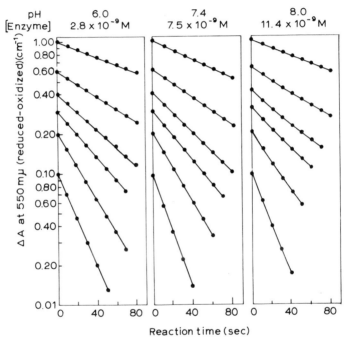

Fig. 2. The kinetics of oxidation of ferrocytochrome c catalyzed by cytochrome oxidase.

The kinetics of oxidation of ferrocytochrome c catalyzed by cytochrome oxidase is approximately first order with respect to the ferrocytochrome c concentration. As shown in fig. 2, the log ferrocytochrome c concentration versus time plot is linear. This was observed by a number of investigators and studied in detail by

Smith and Conrad [14]. Smith and Conrad [14] found that apparent first-order rate constants are a function of the total cytochrome c concentration. In order to explain these anomalies, several somewhat complicated mechanisms have been proposed [15,16]. However, these phenomena may be readily interpreted by assuming the progressive product inhibition by ferrocytochrome c as illustrated in equations (1) and (2) [17].

$$E_{\text{free}} + S \underset{k_2}{\overset{k_1}{\rightleftharpoons}} ES \xrightarrow{k_3} E + P \tag{1}$$

$$E_{\text{free}} + P \underset{k_5}{\overset{k_4}{\rightleftharpoons}} EP \ (ESP \text{ is not formed}) \tag{2}$$

An integrated form of the rate equation for this system, which describes time course of reaction, is:

$$V_{\text{max}} t = [K_{\text{m}} + (p_0 + s_0) K_{\text{m}} / K_{\text{i}}] \ln(s_0 / s)$$

$$+ (1 - K_{\text{m}} / K_{\text{i}})(s_0 - s) . \tag{3}$$

When K_{m} of ferrocytochrome c is approximately equal to K_{i} of ferrocytochrome c, the following equation is obtained:

$$K_{\text{m}} = K_{\text{i}} , \quad V_{\text{max}} t = [K_{\text{m}} + (p_0 + s_0)] \ln(s_0 / s) . \tag{4}$$

Equation (4) describes that the time course of reaction should be first order with respect to the ferrocytochrome c concentration. Rearrangement of equation (4) gives

$$k_{\text{f}} = \ln(s_0 / s) / t = V_{\text{max}} / [K_{\text{m}} + (p_0 + s_0)], \tag{5}$$

which shows clearly that the apparent first order rate constant (k_{f}) is a function of the total cytochrome c concentration: k_{f} decreases as $(p_0 + s_0)$ increases.

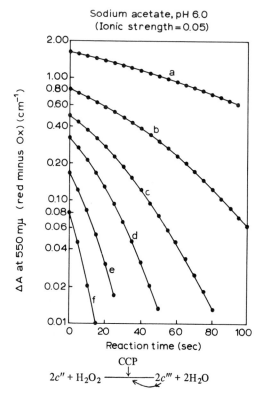

Fig. 3. The kinetics of oxidation of ferrocytochrome c
catalyzed by cytochrome c peroxidase.

The oxidation of cytochrome c catalyzed by cytochrome c peroxidase also obeys the first order kinetics approximately, as reported by Abrams et al. [18], Beetlestone [19], and Nicholls [16].
However, a close examination reveals that the reaction is accelerated with time as shown in fig. 3.

It was found that ferricytochrome c has a substantial amount of
cytochrome c peroxidase activity, so that the overall reaction is
gradually accelerated as ferricytochrome c is produced during the
reaction [20]. This is the reason why this kind of time course curve
is obtained in the cytochrome c peroxidase system. In the presence

of inhibitors, the kinetics of oxidation of ferrocytochrome c catalyzed by cytochrome oxidase or cytochrome c peroxidase often becomes non-first order. Therefore, the apparent first order rate constant, which has been conventionally used, is not a good measure to express the activity of these enzymes. The turnover rate which is based upon the initial steady state rate is a better expression of their activity. The turnover rate of oxidation of ferrocytochrome c varies from 10 per second for horseradish peroxidase, 100 to 1000 per second for cytochrome oxidase to 10,000 per second for cytochrome c peroxidase.

Even since Warburg demonstrated the hemoprotein nature of the terminal oxidase by means of photochemical action spectra [21], the mechanism of action of cytochrome oxidase has been under intensive studies by many investigators. In mitochondria the oxidation of ferrocytochrome c by cytochrome oxidase is coupled by the formation of at least one high energy bond, as demonstrated by Nielsen and Lehninger [22], Chance and Williams [23], and others. The major obstacle in the study of the mechanism of action of cytochrome oxidase is the fact that this enzyme is tightly bound to mitochondrial membranes.

Cytochrome oxidase can be released from mitochondrial membrane by treatment with cholate and salt. This was initially demonstrated by Yakushiji and Okunuki [24] and Straub [25] in 1940. Various methods of purification of cytochrome oxidase devised by many investigators today are still based upon the basic principle of purification laid by these pioneer workers. Detergent-solubilized cytochrome oxidase has been purified to a minimal molecular weight of approximately 100,000 per heme group. Today it is speculated that a basic unit of cytochrome oxidase consists of 1 cytochrome a, 1 cytochrome a_3, and 2 coppers. It is well established that cytochrome a_3 is the closest to oxygen and is sensitive to respiratory inhibitors, such as carbon monoxide and cyanide (cf. fig. 4) The role and reaction sequence of cytochrome a and 2 coppers are yet to be established. Once cytochrome oxidase is

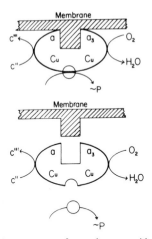

Fig. 4. Components of cytochrome oxidase.

separated from mitochondrial membranes, the phosphorylation coupling mechanism is lost. In addition, there are indications that the link between cytochrome a and cytochrome a_3 is somewhat damaged. For example, in addition of ascorbate and TMPD to a purified cytochrome oxidase, a substantial amount of cytochrome a was reduced, but this partly reduced oxidase does not react with oxygen. This phenomenon was observed by Yakushiji and Okunuki [24] and confirmed in a variety of preparations by subsequent workers [16,26,27]. The reaction sequence of components of the cytochrome oxidase system has been kinetically investigated in membrane-bound cytochrome oxidase by Chance [28]. Gibson and his associates [29] have presented kinetic evidence to show that 1 cytochrome a, 1 cytochrome a_3 and 2 coppers are involved in the reduction of oxygen in detergent-solubilized preparations of cytochrome oxidase. However, a long-standing question whether or not kinetic characteristics of this enzyme in purified preparations really represent a true mechanism of action of cytochrome oxidase in mitochondria still remains until the phosphorylation coupling mechanism in the cytochrome oxidase system is fully reconstituted.

Since there was no indication of technical break-through in sight, we turned our attention to cytochrome c peroxidase a few years ago. In this enzyme system, hydrogen peroxide is the oxidant rather than oxygen. Oxygen carries 4 oxidizing equivalents, whereas hydrogen peroxide carries only 2 equivalents. Thus the latter is considered as a half-reduced oxygen.

Cytochrome c peroxidase was discovered in baker's yeast by Altschul, Abrams, and Hogness in 1940 [12]. These investigators purified this enzyme by a somewhat tedious procedure and demonstrated that this enzyme catalyzes the oxidation of ferrocytochrome c in the presence of hydrogen peroxide. We have shown that cytochrome c peroxidase is an intramitochondrial enzyme like cytochrome oxidase, but this enzyme is liberated from mitochondria upon sonic disruption of mitochondrial structure [30]. We have not been able to detect this enzyme in animal and plant tissues except for aerobically grown yeasts. It is interesting to note that anaerobically grown yeast contains a substantial amount of apocytochrome c peroxidase and the formation of holoenzyme takes place only in the presence of oxygen [31].

By use of DEAE-cellulose chromatography we succeeded in obtaining a highly purified preparation of this enzyme in an excellent yield [31]. Subsequently we found that this enzyme can be crystallized by dialysis against distilled water [33]. Cytochrome c peroxidase has a minimal molecular weight of 35,000 [34]. It is a mono-dispersed monomer containing only one protoheme per molecule. The amino acid composition has been determined [35]. The enzyme is rich in acidic amino acids, such as aspartic acid, threonine, serine, and glutamic acid. Therefore, it has an isoelectric point at pH 5.25. The amino sequence is currently being analyzed by Dr. Matsubara of the University of California. The partial amino acid sequence determined is: NH_2–Thr–Phe–Val ..... Cys ..... Gly–Leu. The enzyme contains one cystein which is not essential for its enzymic activity. PCMB can be attached to this SH group to obtain a heavy atom derivative of the enzyme.

Crystalline cytochrome c peroxidase gives excellent X-ray diffraction patterns, from which the unit cell dimensions were determined to be $57 \times 77 \times 107$ Å [36]. The Z value or the number of molecules per unit cell was 4. Both PCMB and Pt derivatives of the enzyme appeared to be isomorphous to the native enzyme crystal. There is kinetic and physical evidence to indicate the formation of a definite complex between this enzyme and its substrate, cytochrome c. Therefore, we have been trying to co-crystallize the enzyme-substrate complex. The X-ray crystallography of this enzyme is under way in the laboratory of Dr. Kierkegaard at the University of Stockholm.

The protoheme group of cytochrome c peroxidase is non-covalently bound to the apoenzyme moiety. Therefore, the heme and apoenzyme moieties can be reversibly separated in this enzyme. We used Teale's acid-butanone technique to obtain the apoenzyme [35]. Solutions of hemoglobin were adjusted to various pH values and mixed equal volumes of butanone. The mixtures were vigorously shaken and allowed to stand for a minute or so, until butanone and aqueous phases were well separated. Below a certain critical pH value, the heme group was completely separated from the apoprotein and transferred to the upper butanone phase. The heme-free apoprotein remained in the water phase. The heme group of hemoglobin was found to be completely separated below pH 3.5, whereas the heme group of cytochrome c peroxidase was separated below pH 2.5. The so separated apocytochrome c peroxidase has been crystallized by dialysis against distilled water [35]. It is, therefore, possible now to analyse the effect of the heme binding on the conformation of the apoprotein moiety by means of X-ray crystallography.

In order to study the interaction of the apoenzyme with the prosthetic group, we have prepared a variety of modified heme groups, as shown in fig. 5. These modifications are divided into three groups. First, the modification of side chains at positions 2 and 4. Second, the substitution of the central iron with other

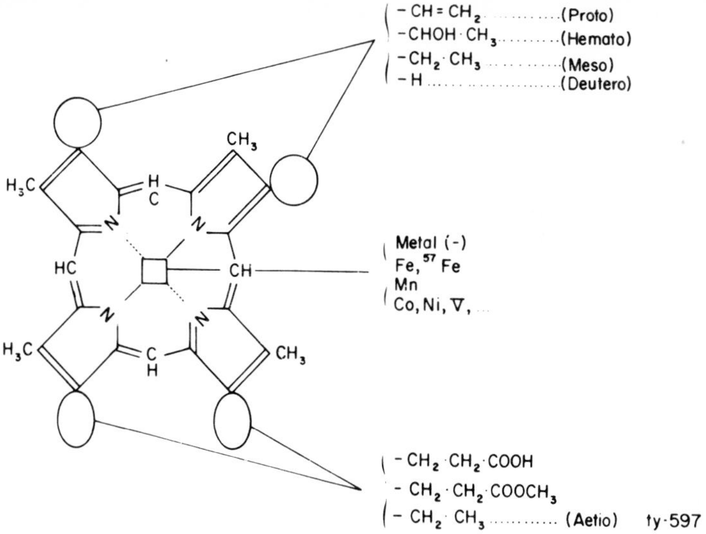

Fig. 5. Various modifications of the prosthetic group of cytochrome *c* peroxidase.

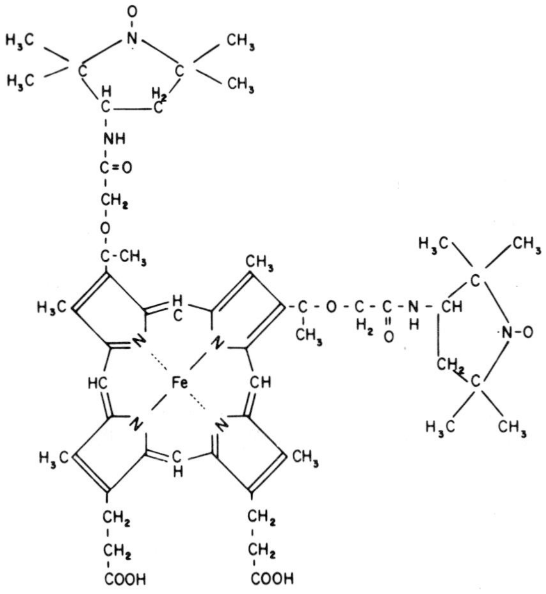

Fig. 6. The structure of 2,4 spin-labeled hemato-heme.

Table 1
Enzymic activities of synthetic cytochrome c peroxidases

CCP containing	Substrate					
	Ferrocytochrome c [a]		Ferrocyanide [b]		Ascorbate [c]	
	V_{max}/e (sec^{-1})	K_m (μM)	V_{max}/e (sec^{-1})	K_m (μM)	V_{max}/e (sec^{-1})	K_m (μM)
Protohemin	5.5×10^3	20	3.2×10^2	18	3.6×10^{-2}	17
Hematohemin	5.5×10^3	25	2.8×10^2	17	2.6×10^{-2}	29
Mesohemin	5.3×10^3	20	3.2×10^2	18	4.0×10^{-2}	23
Deuterohemin	5.2×10^3	28	3.1×10^2	17	3.0×10^{-2}	27
Mn-proto-porphyrin	5.5×10^2	9	3.1×10^1	0.18	1.8×10^{-1}	13

The data for chloroperoxidase are taken from the difference spectra in fig. 5, and also from the absolute spectra on the same samples. For horseradish peroxidase, difference and absolute spectra were made of 1×10^{-5} M enzyme in 0.1 M phosphate buffer, pH 2.8, in the presence and absence of 0.1 M sodium chloride.

transition metals, and thirdly, the modification of side chains at positions 6 and 7.

In the first group, 2 vinyl groups at positions 2 and 4 of protoheme were modified to obtain hemato-, meso-, and deutero-hemes. In addition, as shown in fig. 6, we prepared hematoheme containing 2 spin labels at positions 2 and 4. Table 1 shows the enzymic activity of synthetic cytochrome c peroxidase containing proto-, hemato-, meso-, and deuterohemes. No significant difference was observed [37]. In order to understand the effect of these substitutions on the electronic state of the active center of the enzyme, these synthetic enzymes have been examined by EPR, Mössbauer spectroscopy, and magnetic susceptibility measurements. As shown in fig. 7, EPR spectra of these enzymes were mixtures of high and low spin compounds. No appreciable difference was detected, except for minor difference in the principal g-values.

Light absorption maxima of these enzymes are summarized in table 2. The maxima shift to shorter wave lengths in the order of proto-, hemato-, meso-, and deuterohemes [37]. On the basis of

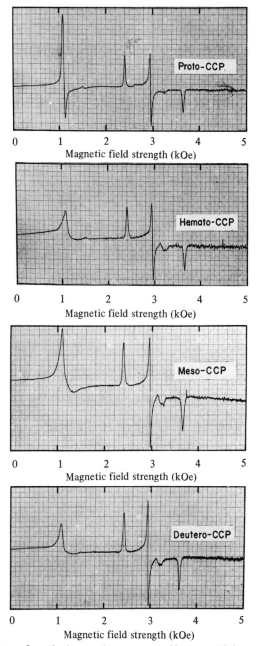

Fig. 7. EPR spectra of synthetic cytochrome *c* peroxidases containing modified hemes.

Table 2
Light absorption maxima of synthetic cytochrome c peroxidases
and their peroxide compounds.
In 0.1 M potassium phosphate buffer, pH 6.0 (23°).

Derivative	Compound	β	CT_1	β	α	CT_2	$\gamma/280$	$\Delta\lambda_\gamma$
Enzyme	CCP	408	505			645	1.25	0
	Hemato-CCP	400	500	–	–	630	1.26	– 8
	Meso-CCP	398	502	–	–	638	1.20	–10
	Deutero-CCP	397	500	–	–	635	1.10	–11
$+ H_2O_2$	CCP	420	–	530	561	625		0
	Hemato-CCP	412	–	525	555	620		– 8
	Meso-CCP	412	–	523	553	615		– 8
	Deutero-CCP	411	–	520	550	610		–10

these results it is reasonable to conclude that side chains at positions 2 and 4 do not play an essential role in cytochrome c peroxidase.

The second group of modification is focussed on the central metal. We found that various porphyrins can recombine specifically with apocytochrome c peroxidase to form well defined porphyrin-protein complexes [38]. These complexes have been crystallized. These complexes had no peroxidase activity, indicating that the iron atom is essential for the enzyme activity, but not for the heme binding to the protein. These porphyrin-apoenzyme complexes were highly fluorescent, so that one can study the energy transfer mechanism between apoprotein and the prosthetic group by fluorescence and phosphorescence measurements. The substitution of the iron atom with other metals such as Co, Mn, Ni, and V has been performed [39]. Co, Ni, and Vanadyl complexes of porphyrins combine with apoenzyme to form well-defined complexes. Although they have no peroxidase activity, these complexes are useful in elucidating the electronic interaction between metalloporphyrin and apoenzyme by physical techniques. All the manganese-porphyrin containing enzymes exhibited peroxidase activity of about 10 percent of the original iron enzyme (cf. table 1).

The third group of modifications is the modification of positions

6 and 7 of heme, where 2 propionic acid side chains are attached in the original protoheme. Etioheme, in which these groups are substituted with 2 ethyl groups, and protoheme dimethyl ester recombine with apoenzyme [40]. It is interesting to note that both the ester and etioheme-protein complexes are soluble even in neutral aqueous buffer solution, although the free forms of these hemes are completely insoluble in the aqueous solutions. This result strongly suggests that the heme is placed in the protein pocket surrounded by hydrophobic amino acid chains.

Table 3
The effect of modification of side chains 6 and 7 heme
on the enzymic activity of cytochrome c peroxidase.

Enzyme	Ferrocytochrome c		Ferrocyanide		Ascorbate	
	V_{max}/e (sec^{-1})	K_m (μM)	V_{max}/e (sec^{-1})	K_m (μM)	V_{max}/e (sec^{-1})	K_m (μM)
CCP	5.5×10^3	20	3.2×10^2	18	3.6×10^{-2}	17
Ester-CCP	2.3×10^2	202	2.2×10^2	18	4.6×10^{-1}	21
Etio-CCP	5.3×10^2	278	1.7×10^2	18	4.1×10^{-1}	25

Table 3 shows the enzymic activity of these artificial enzymes. The activity toward ferrocytochrome c was substantially reduced upon these modifications. As shown in fig. 8, when side chains at positions 6 and 7 were extended from dimethyl ester to dipentylester, the dissociation constant between heme and apoenzyme decreased proportionally. Here the ratio of the bound heme to apoenzyme may be estimated from the ratio of Soret band to 280 mμ protein band. However, the specific activity of the enzyme per bound heme appears to be independent of the length of these side chains, as shown in table 4. From these results it is concluded that there is more than one kind of linkage between heme and protein: one of them is concerned with the iron atom and the other with the carboxyl groups of heme. The metal atom is essential for the

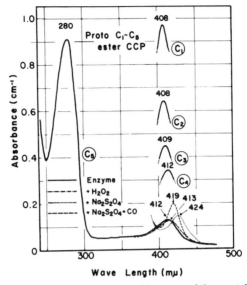

Fig. 8. Absorption spectra of cytochrome *c* peroxidases containing protoheme dialkylesters. The numerical figures indicate the lengths of alkyl groups.

peroxidase activity but not always necessary to combine with the apoprotein when carboxyl groups of heme are free. On the other hand, carboxyl groups are essential to combine with apoprotein when iron atom is absent, but not essential when metal is present

Table 4
The effect of the length of side chains 6 and 7 of protoheme
in enzymic activity of cytochrome *c* peroxidases.

Enzyme containing	Ferrocytochrome *c*		Ferrocyanide	
	V_{max}/heme (sec^{-1})	K_m (μM)	V_{max}/heme (sec^{-1})	K_m (mM)
Protohemin	5500	20	320	18
Protohemin dimethylester (C₁)	13	11	130	4.5
Protohemin diethylester (C₂)	12	9	80	4.2
Protohemin dipropylester (C₃)	33	12	25	3.9
Protohemin dibutylester (C₄)	19	11	16	3.5
Protohemin dipentylester (C₅	26	14	6	1.7

in the porphyrin-ring. As to the enzyme activity, iron atom or other metals are essential for the peroxidase activity. Carboxyl groups of heme are not essential for the enzyme activity, but they play an important role to express full activity.

These results have been further confirmed by using spin-labeled heme derivatives; we attached them to the side chains of the porphyrin ring. By reacting these SL-hemes with apo-CCP, we could obtain SL-CCP of which the hemes have SL compound. The EPR spectra of the SL-CCP were quite different depending on the position of the spin labeling to the heme side chains [41].

Using these natural and synthetic cytochrome *c* peroxidases we investigated the mechanism of action of cytochrome *c* peroxidase. As demonstrated by Keilin [42], Theorell [43], and Chance [44], hydroperoxidases, to which cytochrome *c* peroxidase belongs, form enzyme-substrate compounds with hydrogen peroxide. For example, on the addition of hydrogen peroxide, horseradish peroxidase is converted to green compound I. Compound I is then gradually converted to red compound II, which spontaneously decomposes on standing. The kinetics of formation and decomposition of compounds I and II in the presence of reducing agents are illustrated in fig. 9 [45]. When hydrogen peroxide is added to a mixture of horseradish peroxidase and reducing agents such as ferrocyanide and ferrocytochrome *c*, the brown enzyme is first converted to green compound I. Compound I so formed is rapidly converted to red compound II with a concomitant oxidation of the reducing agent. Compound II reacts slowly with another molecule of the reducing agent to form the original brown enzyme. Therefore, the reaction mechanism shown in fig. 9 was proposed by Chance [45] in 1952.

In the absence of an added reducing agent, these conversions of horseradish peroxidase to compounds I and II take place at a much slower time scale (cf. fig. 10 [46]). The formation of compound I upon the addition of a stoichiometric amount of hydrogen peroxide is indicated by absorbance decrease at 411 mμ. The formed com-

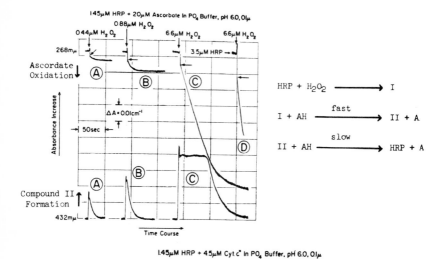

$$HRP + H_2O_2 \longrightarrow I$$

$$I + AH \xrightarrow{\text{fast}} II + A$$

$$II + AH \xrightarrow{\text{slow}} HRP + A$$

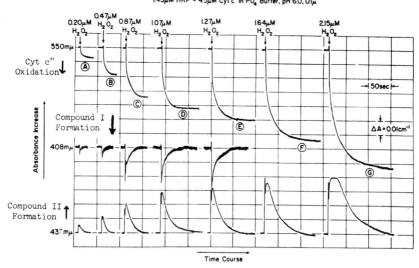

Fig. 9. The kinetics of formation and decomposition of peroxide compounds of horse-radish peroxidase in the presence of ferrocyanide and ferrocytochrome c.

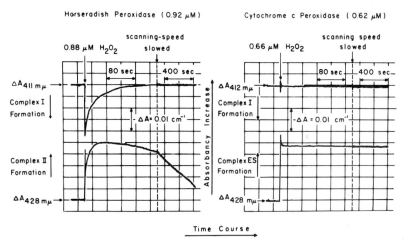

Horseradish Peroxidase (0.92 μM)   Cytochrome c Peroxidase (0.62 μM)

Fig. 10. The kinetics of formation and decomposition of peroxide compounds of horseradish and cytochrome c peroxidases in the absence of a reducing agent.

pound I gradually decomposes with a concomitant formation of compound II as indicated by the 428 mμ absorbance increase. Both compounds I and II decomposed to form the original enzyme. The kinetics of cytochrome c peroxidase is shown on the right. On the addition of hydrogen peroxide the enzyme was rapidly converted to a red compound. The green compound I was apparently not detected under these conditions. The red compound was initially called compound II, since its absorption spectrum is almost identical with compound II of horseradish peroxidase. However, we have called it complex ES on the basis of the new findings described below. Complex ES of cytochrome c peroxidase is highly stable in the absence of the second substrate, ferrocytochrome c. Brown crystals of the enzyme may be converted to red crystals of complex ES upon addition of peroxide [46]. No destruction of crystalline structure was involved in this reaction. Therefore, it is feasible to perform the X-ray analysis of an enzyme intermediate. The spectrum of complex ES, which has maxima at 420, 530, and 561 mμ, is very similar to that of compound II of horseradish peroxidase, so that it is reasonable to assume that the electronic states of the heme iron in these two peroxide intermediates are similar. Since the red compound II of horseradish peroxidase was shown to retain one oxidizing equivalent, the structure of the heme iron in these intermediates was assumed to be a ferryl or Fe^{IV} state. However, the titration of complex ES with reducing agents such as ferrocyanide and ferrocytochrome c showed that complex ES retains two oxidizing equivalents per molecule [45,47]. This suggests that one of the two equivalents must be in somewhere other than the heme iron.

EPR examinations of complex ES revealed that this red compound contained a free radical. Fig. 11 illustrates EPR spectra of the enzyme and complex ES at liquid nitrogen temperature. The EPR signals of the enzyme of high and low spin ferric iron disappeared upon the formation of complex ES [48]. An intense EPR signal of a free radical type was detected at a g value of 2,004. As shown in fig. 12 [37], synthetic cytochrome c peroxidases containing modified heme groups also formed stable complex ES containing free radicals. When complex ES was titrated by successive additions of ferrocytochrome c, the free radical signal was gradually replaced by the EPR signal of the original enzyme [48]. It was shown that one mole of cytochrome c peroxidase is converted to 1 mole of complex ES on the addition of 1 mole of hydrogen peroxide and that 2 moles of ferrocytochrome c are required to convert complex ES to the original enzyme. Therefore, the molar stoichiometry between the enzyme and hydrogen peroxide is 1:1. The stoichiometry between ES and ferrocytochrome c is 1:2. Therefore, the oxidation state of complex ES is formally 2 equivalents higher than ferric. Complex ES contains a free radical and its magnetic susceptibility was determined to be 4.2 Bohr magnetons. On the base of these findings, a structure containing Fe^{IV} and a free radical, X^*, has been proposed for complex ES. This observation serves as the first indication that the apoprotein moiety may be directly involved in the transfer of oxidizing equivalents in the

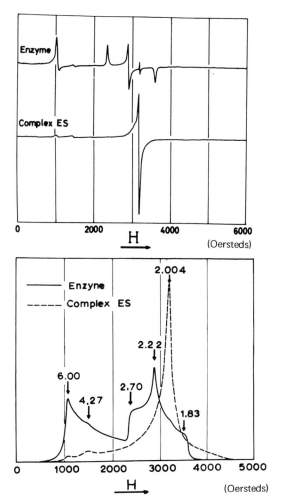

Fig. 11. EPR spectra of cytochrome *c* peroxidase and complex ES at liquid nitrogen temperature.

peroxidase-catalyzed reactions. It is extremely important to elucidate the mechanism by which two presumable unstable species, namely, a ferryl ion and a free radical, can coexist in a stable state. We are trying to identify the chemical nature of the free radical forming group in this enzyme.

44

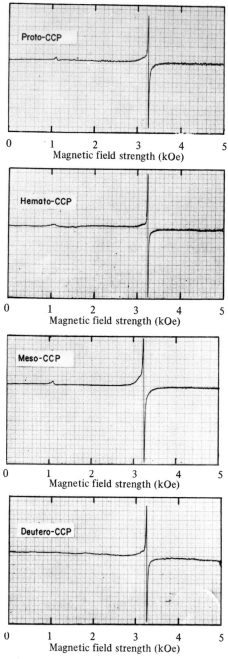

Fig. 12. EPR spectra of complex ES of synthetic cytochrome c peroxidases containing modified hemes.

A hydroperoxidase is a ferric hemoprotein of a brown color. It reacts with a stoichiometric amount of hydroperoxide to form a green compound, compound I, which retains two oxidizing equivalents. Since complex ES of the yeast peroxidase retains also two equivalents, the conversion of the green compound I to the red complex ES must be an intramolecular dismutation. The red compound II retains only one equivalent. Thus, the conversion of the red complex ES to the red complex II should be a one-equivalent reduction, keeping the electronic state of the heme group unchanged. The conversion of the red complex II to the original enzyme of a brown color is a one-equivalent reduction involving the reduction of a ferryl ion to a ferric one. In fig. 13, the overall reactions of horseradish peroxidase and yeast cytochrome *c* peroxidase are compared [46]. One cycle of the horseradish peroxidase reaction consists of the formation of the green compound I, followed by two steps of one-equivalent reduction of hydrogen donors involving the formation and reduction of the red compound II. The corresponding cycle for cytochrome *c* peroxidase reaction

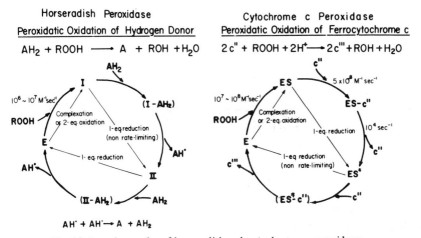

Fig. 13. Reaction cycles of horseradish and cytochrome *c* peroxidases.

is shown on the right. The enzyme is converted to the red complex ES with a second order rate constant of 10^8 M^{-1} sec^{-1}. Complex ES reacts with ferrocytochrome c very rapidly. We would like to mention here that the rate constant between complex ES and ferrocytochrome c was estimated to be 5×10^8 M^{-1} sec^{-1}. This is the most rapid protein–protein interaction that has as yet been measured in solution. This rate constant closely approaches the limits set out by collisions of the two proteins.

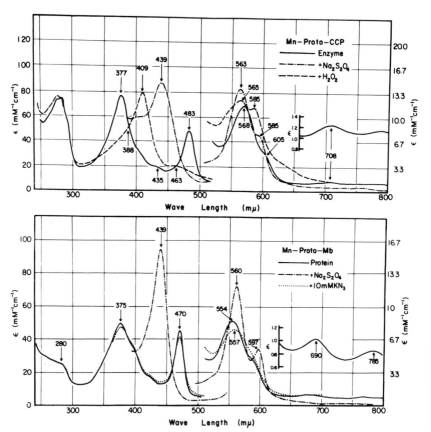

Fig. 14. Absorption spectra of manganese-protoporphyrin-containing cytochrome c peroxidase and myoglobin and their derivatives.

Finally, we would like to discuss manganese-porphyrin-containing enzymes [37,39]. All the manganese porphyrin-protein complexes thus far prepared contain manganese IV in a resting state (cf. fig. 14). On reduction with dithionite, manganese is reduced to a II state. Manganese cytochrome *c* peroxidase and manganese-horseradish peroxidase can be oxidized to a manganese IV state upon the addition of a stoichiometric amount of hydrogen peroxide. However, manganese-containing myoglobin could not be oxidized under these conditions. This indicates that the oxidation-reduction property of manganese porphyrin is highly dependent on the nature of protein bound. As shown in fig. 15, the peroxide compound of manganese-cytochrome *c* peroxidase exhibited an intense free radical signal,

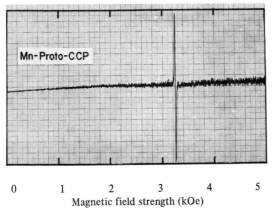

Fig. 15. EPR spectrum of the peroxide compound of manganese-protoporphyrin-cytochrome *c* peroxidase at 77°K.

whereas the peroxide compound of manganese-horseradish peroxidase did not. Therefore, it is likely that the free radical may be associated with the apocytochrome *c* peroxidase moiety.

References

[1] D.Keilin, Ergebn. Enzymforsch. 2 (1933) 239.
[2] D.Keilin, Proc. Roy. Soc. (London) B106 (1930) 418.
[3] H.Theorell and A.Akeson, J. Am. Chem. Soc. 63 (1941).
[4] E.Smith, in: Structure and Function of Cytochromes, eds. K.Okunuki, M.D.Kamen and I.Sekuzu (University of Tokyo Press, Tokyo, 1968) p. 282.
[5] E.Margoliash, in: Hemes and Hemoproteins, eds. B.Chance, R.W.Estabrook and T.Yonetani (Academic Press, New York, 1966) p. 371.
[6] R.E.Dickerson, M.L.Kopka, J.E.Weinzierl, J.C.Varnum, D.Eisenberg and E.Margoliash, in: Structure and Function of Cytochromes, eds. K.Okunuki, M.D.Kamen and I.Sekuzu (University of Tokyo Press, Tokyo, 1968) p. 225.
[7] S.Sano, Proc. Japan. Biochem. Soc. (1968), in press; cf. also in: Structure and Function of Cytochromes, eds. K.Okunuki, M.D.Kamen and I.Sekuzu (University of Tokyo Press, Tokyo, 1968) p. 370.
[8] B.Chance, Biochem. J. 103 (1967) 1.
[9] H.R.Drott, Federation Proc. 28 (1969), in press.
[10] C.P.Lee and L.Ernster, in: Regulation of Metabolic Processes in Mitochondria, eds. J.M.Tager, S.Papa, E.Quagliariello and E.C.Slater (Elsevier, Amsterdam, 1967) p.218.
[11] D.Keilin and E.F.Hartree, Proc. Roy. Soc. (London) B127 (1939) 167.
[12] A.M.Altschul, R.Abrams and T.R.Hogness, J. Biol. Chem. 136 (1940) 777.
[13] B.Chance, in: Enzymes and Enzyme Systems, ed. J.T.Edsall (Harvard University Press, Cambridge, 1951) p. 93.
[14] L.Smith and H.Conrad, Arch. Biochem. Biophys. 63 (1956) 403.
[15] K.Minnaert, Biochim. Biophys. Acta 50 (1961) 23.
[16] P.Nicholls, Arch. Biochem. Biophys. 106 (1964) 25.
[17] T.Yonetani and G.S.Ray, J. Biol. Chem. 240 (1965) 3392.
[18] R.Abrams, A.M.Sltschul and T.R.Hogness, J. Biol. Chem. 142 (1942) 303.
[19] J.Beetlestone, Arch. Biochem. Biophys. 89 (1960) 35.
[20] E.Mochan and B.S.Kabel, Federation Proc. 28 (1969), in press.
[21] O.Warburg and E.Negelein, Biochem. Z. 214 (1929) 64.
[22] S.O.Nielsen and A.L.Lehninger, J. Am. Chem. Soc. 76 (1954) 3860.
[23] B.Chance and G.R.Williams, Advan. Enzymol. 17 (1956) 65.
[24] E.Yakushiji and K.Okunuki, Proc. Imp. Acad. Tokyo 16 (1940) 299; 17 (1941) 38.
[25] F.B.Straub, Hoppe-Seyler's Z. Physiol. Chem. 268 (1941) 227.
[26] T.Yonetani, J. Biol. Chem. 235 (1960) 3138.
[27] T.Yonetani, in: Intracellular Respiration, ed. E.C.Slater (Pergamon Press, London, 1963) p. 396.
[28] B.Chance, Nature 169 (1952) 215.
[29] Q.H.Gibson and D.C.Wharton, in: Structure and Function of Cytochromes, eds. K.Okunuki, M.D.Kamen and I.Sekuzu (University of Tokyo Press, 1968) p. 5.
[30] T.Yonetani and T.Ohnishi, J. Biol. Chem. 241 (1965) 2983.
[31] A.A.Sels and C.Cocriamont, Biochem. Biophys. Res. Commun. 32 (1968) 192.
[32] T.Yonetani and G.S.Ray, J. Biol. Chem. 240 (1965) 4503.
[33] T.Yonetani, B.Chance and S.Kajiwara, J. Biol. Chem. 241 (1966) 2981.
[34] N.Ellfork, Acta Chem. Scand. 21 (1967) 1921.

[35] T.Yonetani, J. Biol. Chem. 242 (1967) 5008.

[36] P.Kierkegaard, L.Hagman and L.Larsson, personal communication.

[37] T.Yonetani and T.Asakura, J. Biol. Chem. 243 (1968) 3996, 4715.

[38] T.Asakura and T.Yonetani, J. Biol. Chem. 244 (1969).

[39] T.Yonetani and T.Asakura, Federation Proc. 27 (1968) 256.

[40] T.Asakura and T.Yonetani, Federation Proc. 28 (1969), in press.

[41] T.Asakura, H.R.Drott and T.Yonetani, in preparation.

[42] D.Keilin and T.Mann, Proc. Roy. Soc. (London) B122 (1937) 119.

[43] H.Theorell, Enzymologia 10 (1941) 250.

[44] B.Chance, J. Biol. Chem. 151 (1943) 553.

[45] T.Yonetani, J. Biol. Chem. 241 (1966) 2562.

[46] T.Yonetani, H.Schleyer, B.Chance and A.Ehrenberg, in: Hemes and Hemoproteins, eds. B.Chance, R.W.Estabrook and T.Yonetani (Academic Press, New York, 1966) p. 293.

[47] T.Yonetani, J. Biol. Chem. 24 (1965) 4509.

[48] T.Yonetani, H.Schleyer and A.Ehrenberg, J. Biol. Chem. 241 (1966) 3240.

THE ROLE OF MAMMALIAN PEROXIDASE
IN IODINATION REACTIONS

M. MORRISON, G. BAYSE and D. J. DANNER

*Department of Biochemistry, St. Jude Children's Research Hospital,
332 N. Lauderdale, P.O. Box 318, Memphis, Tennessee 38101, USA*

1. INTRODUCTION

The peroxidases, a widely distributed group of enzymes, are
found in fungi and bacteria as well as in all mammals and all higher
plants [1−3]. Most reviews of the subject suggest that a common
reaction mechanism exists for oxidations catalyzed by all peroxidases, and further, that all peroxidases participate in the same types
of reactions. The objectives of this article will be to review the enzymes in mammalian tissues and to point out the differences between peroxidases.

2. LOCALIZATION

The localization of peroxidases in mammalian tissues is not a
simple problem [4]. Peroxidase activity is ubiquitous to all mam-

malian tissues. This is attributable to the fact that pseudoperoxidases such as hemoproteins, copper proteins, etc. will show peroxidase activity. The true peroxidase is distinguished from the so-called pseudoperoxidase by a much higher specific activity. In tissue localization, the difficulty in distinguishing true peroxidase from pseudoperoxidase activity has often led to erroneous interpretations. Unequivocal localization, therefore, has depended upon isolation or a specific assay for the protein. Thus, immunochemical assessment is necessary for the specific localization of the mammalian peroxidase. In the case of lactoperoxidase, the purified enzyme was used as an antigen, and antiserum against the enzyme was used to screen tissues [4,5]. With this technique, it was possible to localize the lactoperoxidase in the salivary, lacrimal, harderian, and mammary glands, but not in other tissues. In a similar manner, thyroid peroxidase was localized only in the thyroid. Thus, in the case of mammalian tissues, the hemoprotein peroxidases reflect the specialized role of the tissues in which they are found.

Besides lactoperoxidase, three other peroxidases have been isolated, purified, and clearly distinguished in the mammalian tissues. Myeloperoxidase is present in the white cell series and thyroid peroxidase is present in the thyroid gland. Glutathione peroxidase, a flavoprotein, is widely distributed, being present in the mitochondria and even in the mature red cells which contain no mitochondria.

Not all the cells of each of these tissues contain peroxidase. In the case of white cells, the lymphocytes do not contain myeloperoxidase. In the salivary glands, it was shown that the acinar cells, but not the duct cells, contain lactoperoxidase [6]. Furthermore, not all of the salivary glands appear to contain this enzyme. In bovine tissue, the submaxillary and sublingual glands contain the enzyme, while the parotid does not. In the pig, however, the parotid does contain the enzyme, while the lacrimal is deficient in lactoperoxidase [6].

Table 1
Hemoprotein content of pig thyroid microsomal fractions *

Hemoprotein	μmoles/mg protein
Cytochrome b_5	0.033
P 450	0.000
Peroxidase	0.066

* Hosoya and Morrison, Biochem. 6 (1967) 1021.

Thyroid peroxidase is present in the microsomal fraction of the thyroid tissue [7]. The microsomes contain no P450 but do contain cytochrome b_5, the peroxidase and a non-heme iron protein. Thus, it would appear that P 450 is absent and is replaced by a peroxidase in the thyroid microsomes and that the microsomal oxidation system is modified for specialized function of the tissue.

3. PHYSICAL, CHEMICAL STUDIES OF PEROXIDASES

The peroxidases present an interesting group of enzymes from the physiochemical point of view. Spectral properties of their prosthetic groups make them easily studied in the visible and ultraviolet regions of the spectrum. They can also be studied by EPR, NMR and Mössbauer and, therefore, lend themselves well to studies of structure-function relationships. A comparison of the spectral properties of peroxidases clearly points out that the peroxidases in mammalian tissues are not isozymes but distinctly different proteins. Further, the heme prosthetic groups, even though their structures are unknown, are clearly different; and finally, the protein ligands of the various peroxidases are quite different.

Lactoperoxidase is structurally interesting since its heme prosthetic group is linked covalently to the protein [18]. This bond is not the thio ether of cytochrome c, but is an ester linkage involving a carboxyl group of the protein and a hydroxyl group of the

Table 2
Properties of various peroxidases

	Mol. wt.	Heme/ mole	Maximum soret peak				References
			Oxidized	Reduced	Reduced-CO	CN$^-$	
Lactoperoxidase	78,000	1	412	438	425	432	[8,9,10]
Thyroid peroxidase	160,000	3	412	425	420	428	[11,12]
Myeloperoxidase	149,000	2	430	475			[13,14]
Horseradish peroxidase	40,000	1	403	440	423	432	[15]
Chloroperoxidase	42,000	1	400	409		450	[16]
Cytochrome c peroxidase (yeast)	49,000	1	408	438	423	424	[17]

heme. The actual structure of the heme is not fully established, but reductive cleavage of the heme from the protein gives meso-porphyrin IX. This establishes that the heme is directly related to the widely-distributed protoporphyrin IX.

The amino acid composition of lactoperoxidase is not unique, but it does have a high carbohydrate content [10]. This is true of all the peroxidases yet studied, with the exception of the yeast cytochrome c peroxidase which may represent a variant type. Lactoperoxidase contains something on the order of 38 moles of carbohydrate per mole of enzyme. It has been suggested that carbohydrate may represent the endogenous electron donor of the peroxidase [19].

4. ACTIVE CENTER OF LACTOPEROXIDASE

The spectral properties of lactoperoxidases as a function of pH provide some information as to the nature of the groups which are coordinated to the prosthetic group.

At low pH values, there is a shift in spectra between pH 3 and 5 which is completely reversible. This suggests that the group ionizing with this pK is a carboxyl group, probably that of aspartic acid. At high pH values, a group ionizes with a pK at 11.2 and probably is the coordinating group involved in the spectral shift. This group may be either the guanidine of arginine, or the epsilon amino of lysine.

A modification of the above approach is a study of ligand exchange in hemoproteins. Orgel [20] has suggested the order of ligand complexing as $CN > NH_2 > H_2O > F > Cl > Br > I$. A study of these ligands shows that lactoperoxidase forms a complex with cyanide. The difference spectra are shown in fig. 1. Since the greatest differences occur at 432 nm, the rate of the reaction between cyanide and lactoperoxidase was studied at that wavelength [9]. This is a rapid reaction and the kinetics were, therefore,

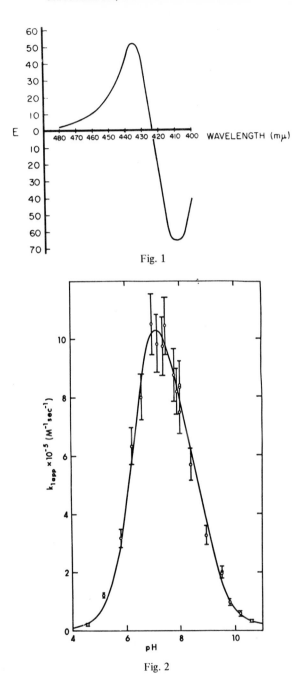

Fig. 1

Fig. 2

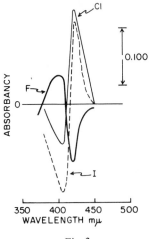

Fig. 3

studied with relaxation methods in collaboration with Dr. Brian Dunford and his colleagues. The first order kinetics of the relaxation process were followed as a function of pH and are shown in fig. 2.

These results can most simply be explained by the presence of ionizable groups on the enzyme which affect the rate of cyanide complex formation. The simplest mechanism that will explain the data satisfactorily involves two ionizable groups. The pK values for these ionizable groups are 6.3 and 7.6. Neglecting possible differences between molecular and group ionization constants, these pK values fall into the range of the pK attributable to the imidazole group of histidine, and the alpha amino group of an amino acid residue. Since our amino acid analysis showed only a single amino terminal leucine, the heme group may be situated toward the N-terminal end of the molecule. Since neither of these two groups are heme ligands, they must be in a pathway between the outside of the protein and the heme prosthetic group. Thus, their ionization affects the kinetics of the cyanide complex formation.

The halides also form complexes with the enzyme. Fig. 3 shows

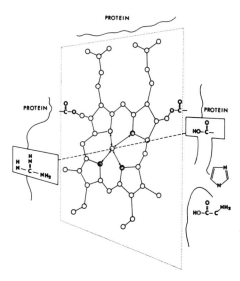

Fig. 4. Proposed structure for prosthetic group of lactoperoxidase

the difference spectra of the halide complexes with lactoperoxidase. Most interesting is the fact that fluoride shifts the maxima
to lower wavelengths, suggesting a shift to higher spin state while
chloride, bromide and iodide, like cyanide, shift the maxima to
longer wavelengths indicating a shift to lower spin state. These results are consistent with known electronegativity of the halides.

The data on the active site of lactoperoxidase is summarized in
fig. 4. The prosthetic group is linked to the protein via ester linkage and has a carboxyl and a guanidinium group as ligands. An
imidazole of the protein and an alpha amino are in the intramolecular pathway of cyanide to the heme, which is located in the
N-terminal region of the peptide chain. Lactoperoxidase at 22°
and neutral pH values is in a mixed spin state and is shifted to a
higher spin state when complexed with fluoride, and a lower spin
state when complexed with cyanide, iodide, bromide or chloride.

5. CATALYTIC ACTIVITY OF PEROXIDASES

The enzyme activity of peroxidases can be assayed in a variety of ways although their ability to catalyze the oxidation of aromatic phenols or amines has been the most common method employed. The colored oxidation products have high extinctions which are readily evaluated by spectrophotometric analysis [21]. The specific activity of peroxidases, based on the heme prosthetic group, has the same order of magnitude for all of the peroxidases with these electron donors as shown in table 3.

Table 3
Comparison of the ability of hemoprotein to catalyze the oxidation of guaiacol

Hemoprotein preparation	GU/μmole *
Lactoperoxidase	4.40
Myeloperoxidase	2.10
Purified thyroid peroxidase	1.15
Horseradish peroxidase c	1.50
Horseradish peroxidase	2.20
Horse hemoglobin	0.000006
Human hemoglobin	0.000045
Horse heart cytochrome c	0.000017

* Assay and units are described in Hosoya and Morrison, J. Biol. Chem. 242 (1967) 2828.

Peroxidase can participate in iodination reactions, and equations (a–c) have been proposed for the reactions involved.

a. $H_2O_2 + I^- \rightarrow$ "oxidized I"
$I^- \rightarrow I^{\cdot} + e$
$I^- \rightarrow I^+ + 2e$

b. $2I^{\cdot} + I^- \rightarrow I_3^-$
$I^+ + 2I^- \rightarrow I_3^-$

c. "oxidized I" + $HO-\langle O \rangle - R \rightarrow HO-\overset{I}{\langle O \rangle} - R$

The peroxidase is involved in the oxidation of iodide as illustrated in reaction a, although the mechanism of oxidation is unknown. Iodide may be oxidized by a one or a two electron transfer process and, in either case, I_3^- would be produced (equation b). In the iodination reactions, it has been suggested that a second enzyme, an iodinase [22,23] is involved. This enzyme supposedly catalyzes the reaction between the oxidized iodide and the phenolic compound (quotation c). The role of the peroxidase in iodination, according to this scheme, is simply to oxidize iodide. Therefore, an assay for the ability of peroxidases to oxidize iodide has been employed by investigators in this field. This is a spectrophotometric method [24−26] in which the oxidation product of iodide , I_3^- , is readily detected in the ultraviolet. To investigate the overall reaction, the most widely used procedure for the study of halogenation reactions catalyzed by peroxidases is a chemical assay of the incorporation of the halide into an organic molecule [27]. In order to increase the sensitivity of this type of procedure, radioactive isotopes of the halides are usually employed.

Another method of following halogenation reactions involves the determination of a change in the spectral properties of a molecule. Hager and his co-workers [28] have devised a spectrophotometric procedure appropriate for kinetic studies.

Table 4
Specificity of iodination assay

	ΔI^- (μmoles/min/mg)
Lactoperoxidase	240.00
Thyroid peroxidase	35.10
Myeloperoxidase	6.48
Horseradish peroxidase	0.00
Met-myoglobin	0.00
Hemoglobin	0.00
Cytochrome c	0.00

Table 5
Comparison of iodide oxidation and iodination

	I_3^- production (μmoles/min)	MIT production (μmoles/min)
Lactoperoxidase	0.124	0.270
Horseradish peroxidase	0.139	0.000

A third method of studying halogenation reactions employs the halide specific sensor [29]. With this procedure, concentrations of iodide as low as 1×10^{-7} M can be detected, and the kinetics of iodination can be readily determined by following the rate of change in iodide concentration.

The ability to catalyze the iodination reaction was assayed with this last procedure and found to be much more specific than the oxidation of aromatic amines or phenols. As shown in table 3, all the peroxidases will catalyze the oxidation of guaiacol. However, the pseudoperoxidase will oxidize guaiacol at only about 1/10,000 the rate of the true peroxidase. In the case of iodination, as shown in table 4, lactoperoxidase, myeloperoxidase, and thyroid peroxidase will catalyze the reaction above pH 7.0, but horseradish peroxidase and all the pseudoperoxidases are inactive.

The data in table 5 points out that the ability to oxidize iodide is not equivalent to the ability to catalyze the iodination reaction. In the experiment cited, horseradish peroxidase and lactoperoxidase are matched for their ability to catalyze the oxidation of I^- to I_3^-. Horseradish peroxidase will not catalyze iodination under these conditions, while lactoperoxidase is very effective. Two points are very clear: first, peroxidases can catalyze the iodination reaction in the absence of any other enzyme such as an iodinase, and second, the iodination is not simply attributable to the oxidation of iodide.

With the iodide sensor, a kinetic study of the rate of iodination of a variety of phenolic compounds is possible. Tyrosine is iodinated more readily than MIT, while no iodination of DIT can be

Table 6
Specific activity of lactoperoxidase in iodinating tyrosine

	ΔI (mole/min/mole L.P.)
D-tyrosine	1.2×10^4
L-tyrosine	1.0×10^4
MIT	0.3×10^4
DIT	–

observed. This is what might be expected, based on chemical reactivity of these substrates. More significant was the unexpected finding that D-tyrosine is iodinated more rapidly than L-tyrosine, suggesting that lactoperoxidase is stereospecific with regard to the compound being iodinated. If the carboxyl group of tyrosine is altered by making an amide or ester, the difference in the rate of iodination is not great, but if the phenol group or the alpha amino group is altered, the rate is significantly slower than that of L-tyrosine.

Proteins and peptides are iodinated more slowly than tyrosine. Further, if the tyrosyl residue in the peptides is N-terminal, it is iodinated more readily than if it is carboxyl terminal. The proteins although iodinated at different rates, are all iodinated more readily than the peptides.

Although rates of lactoperoxidase-catalyzed iodination are optimal at pH 5, there is a loss in specificity at low pH values as shown in table 7. Glycyl-L-tyrosine, for example, is iodinated at essentially the same rate at pH 4.5 as L- or D-tyrosine, At pH values of 7.4 and above, the peptide is iodinated at about 1/50 the rate of L-tyrosine. There is also a marked difference in the relative rates of iodination of D- and L-tyrosine at higher pH values. The D-tyrosine is iodinated about 1.5 times more rapidly than L-tyrosine at pH 8.2.

Tyrosine is not the only amino acid residue which can partici-

Table 7
Relative rates of iodination catalyzed by lactoperoxidase

Substrate	pH		
	4.5	7.4	8.2
L-tyrosine	100	100	100
D-tyrosine	107	123	146
Glycyl-L-tyrosine	93	3	2

pate in iodination reactions. Histidine is also iodinated, but at a slower rate than tyrosine. These two amino acids are competitive in the iodination reaction. DIT is also a very effective competitive inhibitor of iodination of tyrosine and of proteins. This competitive inhibition indicates that the substrate which is iodinated must be bound by the enzyme.

Iodination involves what might be thought of as two sites: the heme prosthetic group, and a second site. This second site may be a functional group on the enzyme, or it may simply be a binding site for the compound which is iodinated. In the case of thyroid peroxidase, it is possible to separate the ability to catalyze the iodination reaction while the guaiacol oxidation activity is unaffected. As shown in table 8, when the enzyme is treated with trypsin, there is a marked loss of the ability of the preparation to catalyze iodination but no loss in the ability to catalyze oxidation of guaiacol.

Table 8
Effect of trypsin digestion on thyroid peroxidase activity

	Relative rates	
	Guaiacol oxidation	Iodination
Thyroid peroxidase	100	100
Thyroid peroxidase + trypsin *	100	23

* One hour digest at 25° trypsin/peroxidase (1:10)

Table 9
Stoichiometry of the lactoperoxidase iodination reaction

Substrate	$H_2O_2/\Delta I^-$	$H_2O_2/\Delta MIT$
L-tyrosine	1.0	1.0
L-tyrosylglycine	1.0	
L-histidine	1.7	

Substrate	Iodide incorp. protein (M/M)	$H_2O_2/\Delta I^-$ (M/M)
Lysozyme	0–0.01	2.0
	0.02	1.0
Bovine serum albumin	0–0.02	30.0
	0.25	1.0

The stoichiometry between H_2O_2 consumed and halide incorporated into the amino acids and proteins is shown in table 9. One mole of iodide was incorporated into tyrosine per mole of H_2O_2 consumed. In the early stages of iodination of proteins, however, much more H_2O_2 is consumed. This data indicates that peroxide is being used in reactions which do not result in iodination. The oxidation of groups in the protein unquestionably takes place. Some of the oxidation could occur by reaction with the "oxidized iodide", while other oxidation may take place by direct enzyme action. Lactoperoxidase is capable of oxidizing tyrosine to dityro-

Table 10
Iodination reaction products of thyroglobulin

Enzyme	% of total iodine		
	MIT	DIT	$T_3 + T_4$
Lactoperoxidase	43.4	38.4	18.7
Thyroid peroxidase	50.3	27.4	23.3

syl derivatives in the absence of iodide. It is also capable of oxidizing tryptophan to some unknown derivative in the absence of iodide. Hence, this system can be a complicated one.

We have attempted to evaluate the products of iodination in proteins. In table 10 the products of iodination of thyroglobulin catalyzed by lactoperoxidase and thyroid peroxidase are compared. In these experiments, radioactive iodide was incorporated into the proteins and the products were then separated after the protein had been digested with pronase. The radioactivity incorporated into MIT, DIT, T_3 and T_4 was then measured and the percentage of each of the products formed was evaluated. There is no great difference discernible in products.

Acknowledgments

This work has been supported in part by Public Health Service grants GM 15919, GM 15913, DE 02792 and CA 05176 and by the American Lebanese Syrian Associated Charities (ALSAC).

References

[1] B.C.Saunders, A.G.Holmes-Siedle and B.P.Stark, in: Peroxidase (Butterworths, London, 1964).
[2] R.H.Burris, in: Encyclopedia of Plant Physiology, vol. 12 (Springer, Berlin, 1960) p. 366.
[3] K.G.Paul, in: The Enzymes, eds. P.D.Boyer, H.Lardy and K.Myrback, vol. 8 (Academic Press, New York, 1959) p. 227.
[4] M.Morrison and P.Z.Allen, Biochem. Biophys. Res. Commun. 13 (1963) 490.
[5] M.Morrison, P.Z.Allen, J.Bright and W.Jayasinghe, Arch. Biochem. Biophys. 111 (1965) 126.
[6] M.Morrison and W.F.Steele, in: Biology of the Mouth, ed. P.Person (AAAS, Washington, D.C., 1968) p. 89.
[7] T.Hosoya and M.Morrison, Biochemistry 6 (1967) 1021.
[8] M.Morrison, H.B.Hamilton and E.Stotz, J. Biol. Chem. 228 (1957) 767.
[9] D.Dolman, H.B.Dunford, D.M.Chowdhury and M.Morrison, Biochemistry 7 (1968) 3991.

[10] W.A.Rombauts, W.A.Schroeder and M.Morrison, Biochemistry 6 (1967) 2965.
[11] D.J.Danner and M.Morrison, unpublished observations.
[12] T.Hosoya and M.Morrison, J. Biol. Chem. 242 (1967) 2828.
[13] K.Agner, Acta Chem. Scand. 12 (1958) 89.
[14] J.Shultz and K.Kaminker, Arch. Biochem. Biophys. 96 (1962) 465.
[15] D.Keilin and E.F.Hartree, Biochem. J. 49 (1951) 88.
[16] D.E.Morris and L.P.Hager, J. Biol. Chem. 241 (1966) 1763.
[17] T.Yonetani and G.S.Ray, J. Biol. Chem. 240 (1965) 4503.
[18] D.E.Hultquist and M.Morrison, J. Biol. Chem. 238 (1963) 2843.
[19] M.Morrison, W.A.Rombauts and W.Schroeder, in: Hemes and Hemoproteins, eds.
     B.Chance, R.W.Estabrook and T.Yonetani (Academic Press, New York, 1966)
     p. 345.
[20] L.E.Orgel, in: Haematin Enzymes, vol. 19, eds. J.E.Falk, R.Lemberg and R.K.
     Morton (Pergamon Press, New York, 1961) p. 1.
[21] B.Chance and A.C.Maehly, in: Methods Biochem. Anal., vol. 1, ed. D.Glick (Inter-
     science, New York, 1954) p. 764.
[22] L.J.DeGroot and A.M.Davis, Endocrinology 70 (1962) 492.
[23] C.C.Yip, Biochem. Biophys. Acta 96 (1965) 75.
[24] T.Hosoya, J. Biochem. 53 (1963) 381.
[25] N.M.Alexander and R.Scheig, Anal. Biochem. 22 (1968) 187.
[26] R.P.Igo, C.P.Mahoney and B.Mackler, J. Biol. Chem. 239 (1964) 1893.
[27] J.Roche, S.Lissitzky and R.Michel, in: Methods Biochem. Anal., vol. 1, ed. D.Glick
     (Interscience, New York, 1954) p. 243.
[28] L.P.Hager, D.R.Morris, F.S.Brown and H.Eberwein, J. Biol. Chem. 241 (1966)
     1769.
[29] M.Morrison, Gunma Symposia on Endocrinology 5 (1968) 239.

STUDIES ON THE RELATIONSHIP OF CHLOROPEROXIDASE-HALIDE AND CHLOROPEROXIDASE-HYDROGEN PEROXIDE COMPLEXES TO THE MECHANISM OF THE HALOGENATION REACTION*

Lowell P. HAGER, John A. THOMAS ** and David R. MORRIS ***

Biochemistry Division, University of Illinois, Urbana, Illinois, USA

1. Introduction

Studies from this laboratory have led to the isolation and crystallization of chloroperoxidase from *Caldariomyces fumago* [1]. Chloroperoxidase, in the presence of hydrogen peroxide and an appropriate halogen anion, will catalyze the peroxidative formation of a carbon-halogen bond when supplied with a nucleophilic acceptor molecule [2] according to equation (1).

$$X^- + H_2O_2 + AH \rightarrow AX + OH^- + H_2O \qquad (1)$$

* This research has been supported by a grant (GB 5542) from the National Science Foundation.
** Present address: Johnson Foundation, University of Pennsylvania, Philadelphia, Pennsylvania, USA.
*** Present address: Biochemistry Department, University of Washington, Seattle, Washington, USA.

The halogen donor (X^-) in equation (1) can be chloride, bromide, or iodide ion. Halogen acceptors (AH) in equation (1) can be divided into two groups based on the catalytic rate constant (k_{cat}) for the overall reaction [3]. With group I acceptors, k_{cat} for the overall reaction is independent of the acceptor molecule and depends solely on the halide donor. Thus with chloride ion as the halogen donor, k_{cat} for chlorination with any group I acceptor is 1×10^3. With bromide ion as donor, k_{cat} is 2×10^3 with group I acceptors. With group II acceptors, the rate constant for the overall reaction is variable (5—500) and depends both on the halide donor and the nucleophilic character of the acceptor molecule, however, k_{cat} with group II acceptors is always smaller than that found for a group I acceptor [3].

Chloroperoxidase is a protoheme peroxidase which is very similar to the horseradish and Japanese radish peroxidases in its physical and chemical properties [1]. However, neither horseradish nor Japanese radish peroxidase catalyze chlorination or bromination reactions. Pertinent information concerning the physical and chemical properties of chloroperoxidase is listed in table 1. The spectral properties, purification and partial characterization of chloroperoxidase, have been previously reported [1].

In a study aimed at defining a mechanism for the enzymatic halogenation reaction catalyzed by chloroperoxidase, Brown and Hager [4] compared the isomer ratios resulting from enzymatic chlorination of anisole with chemical chlorination of anisole via both an

Table 1
Properties of chloroperoxidase

1. Prosthetic group: ferriprotoporphyrin IX
2. Minimum molecular weight (heme/dry weight) 42,000
3. Molecular weight (sedimentation): 41,000
4. Carbohydrate content: $\sim 25\%$
5. Principal residues: mannose, glucose, galactose (2:1:1)
6. Turnover number: 66,000 Cl^-, 120,000 Br^-, 250,000 I^-

ionic and free radical route. These studies clearly indicated that the enzyme reaction and the chemical halogenation via an ionic mechanism were very similar if not identical. Both the enzymatic and ionic chemical reactions gave ortho and para monochloroanisoles in a ratio of 1.8−1.9 para to 1 ortho. Free radical chlorination gave quite a different isomer ratio and, in addition, free radical halogenation produced side chain halogenated product (phenoxymethyl chloride) whereas the enzymatic and chemical ionic chlorination did not. These results led to the hypothesis that an enzyme-bound halogenium ion $(Enz-X^+)$ is formed as an active intermediate in the halogenation reaction. It is the purpose of this paper to report on studies directed toward defining reactions which could produce the $Enz-X^+$ intermediate.

2. A model for the chloroperoxidase reaction

The results obtained from transient and steady state kinetic studies and studies on the formation and nature of heme ligands in chloroperoxidase can be best interpreted in terms of a model in which hydrogen peroxide, halide ion, and the acceptor molecule react sequentially with the enzyme in an ordered mechanism. This model is presented in equations (2)–(5).

$$Enz + H_2O_2 \underset{k_{-1}}{\overset{k_1}{\rightleftharpoons}} Enz^{+2} + 2\,OH^- \tag{2}$$

$$Enz^{+2} + X^- \underset{k_{-2}}{\overset{k_2}{\rightleftharpoons}} (Enz - X)^+ \tag{3}$$

$$(Enz - X)^+ + AH \underset{k_{-3}}{\overset{k_3}{\rightleftharpoons}} (Enz - X[AH])^+ \tag{4}$$

$$(Enz - X[AH])^+ \overset{k_4}{\longrightarrow} Enz + AX + H^+ \tag{5}$$

In the first step of this mechanism (equation 2), the binding of hydrogen peroxide results in the formation of an intermediate with two oxidizing equivalents (Enz^{+2}). This reaction is comparable to the formation of horseradish peroxidase complex I (or compound I). The subsequent reactions involve the formation of complexes between the enzyme and halide ion (equation 3) and acceptor (equation 4).

Contributions to the above model have come from extensive steady state kinetic studies by Thomas [5]. His experiments have shown that the halogenation reactions catalyzed by chloroperoxidase yield families of parallel lines in double reciprocal plots of initial velocity against the concentration of one substrate at several fixed concentrations of the other substrate. In the terminology of Cleland, this type of kinetic behavior is considered to be evidence for a ping pong bi bi mechanism, with binary enzyme kinetics [6]. The results of the steady state kinetic studies impose two conditions on any mechanism proposed for chloroperoxidase: (1) substrates must be added in an obligatory order to the enzyme (2) the first step is essentially irreversible either because the binding of the first substrate is followed by the release of a product before the second substrate is bound, or the substrate dissociation constants are low compared to the substrate concentration used.

The experimental results which follow are primarily addressed to the steps outlined in equations 2 and 3 (steps 1 and 2) in the model. Spectral evidence has been obtained for the formation of complexes (or compounds) between chloroperoxidase and hydrogen peroxide and chloroperoxidase and halides. The kinetics and pH dependencies of complex formation can be correlated with the properties of the overall reaction and a consistent fit between the model and experimental result can be found.

The first step in the ordered sequence of reactions between chloroperoxidase and its substrates must be a reaction between the enzyme and hydrogen peroxide. As indicated above, steady state kinetics require that the first step either be essentially irreversible or be accompanied by the release of a product. Neither halide ion nor the halogen acceptor fulfill these conditions; however, as will be seen, the reaction between hydrogen peroxide and chloroperoxidase does meet these conditions.

3. Step I. Formation of enzyme-hydrogen peroxide complex I

The addition of hydrogen peroxide to chloroperoxidase results in the rapid formation of an intermediate having spectral properties similar to horseradish complex I [7] (fig. 1, curve I). For comparative purposes, the spectrum of native chloroperoxidase (CPO) is included in fig. 1. The spectrum of the initial peroxide complex is characterized by a general loss of absorbance in the Soret band. The wavelength of maximal absorption of the complex is about 400 mμ, as compared with 407 mμ for horseradish peroxidase. The spectrum and the rate of formation of the CPO-hydrogen peroxide

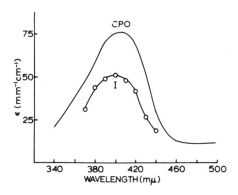

Fig. 1. *Spectrum of chloroperoxidase-hydrogen peroxide complex I.* The spectrum of the initial peroxide complex I was determined pointwise in a dual wavelength stopped-flow spectrophotometer. The enzyme and peroxide concentrations after mixing were 1.4×10^{-6} M and 1.5×10^{-5} M respectively. The reference wavelength was 450 mμ. The spectrum of complex I was constructed from the change in transmittance between time zero and one second after mixing. The solutions were buffered at pH 2.8 with 0.1 M potassium phosphate. The spectrum of native enzyme (curve labeled CPO) is provided for comparison.

complex are the same at pH 3 and 6. This is an important point which will enter into the later discussion.

4. Rate of formation of complex I

The rate of formation of complex I has been determined by the rate of change of absorbance at 410 mμ when enzyme and hydrogen peroxide are mixed in the Gibson-Durham stopped flow instrument. Fig. 2 records an oscilloscope tracing of the time course of the absorbance change. The second order rate constant was calculated from the initial rate by use of the simplified equation for a pseudo first order reaction [8] according to equation (6) where E_0 is the initial enzyme concentration,

$$k_1 = \frac{2.3}{S_0 t} \log \frac{(E_0)}{(E_0 - ES)} \tag{6}$$

ES is the concentration of the enzyme-peroxide complex at time = t, and S_0 is the initial peroxide concentration. The concentration of ES was assumed to be equal to E_0 when the reaction was over.

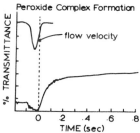

Fig. 2. *Rate of formation of chloroperoxidase-hydrogen peroxide complex I.* Chloroperoxidase and hydrogen peroxide were mixed in 0.1 M potassium phosphate buffer, pH 2.8, in a dual-wavelength stopped-flow apparatus. The light intensities were measured on a Techtronix type 564 storage oscilloscope and recorded with a Polaroid camera. One syringe of the stopped-flow apparatus contained enzyme and phosphate buffer, the other syringe contained hydrogen peroxide. After mixing, the enzyme concentration was 1.4×10^{-6} M and the hydrogen peroxide concentration was 1.3×10^{-5} M. The reaction chamber has a 1 cm light path. It was monitored at 410 mμ with 450 mμ as the reference wavelength.

Table 2

The rate constant for the formation of chloroperoxidase-hydrogen peroxide complex I.

Concentrations		k_1 (liters moles^{-1} sec^{-1})
Enzyme	Hydrogen peroxide	
1.5×10^{-6} M	1.3×10^{-5} M	0.9×10^6
1.5×10^{-6} M	6.5×10^{-6} M	1.6×10^6
1.5×10^{-6} M	3.2×10^{-6} M	1.6×10^6
3.0×10^{-6} M	1.0×10^{-5} M	1.5×10^6

The rate of complex I formation was measured in the stopped flow spectrophotometer as described in fig. 2. All the reactions were carried out at pH 2.8 in 0.1 potassium phosphate buffer. The rate constants were calculated by the method of Chance [8].

Table 2 shows the values for k_1 obtained at different concentrations of enzyme and hydrogen peroxide. Clearly, the formation of complex I is a very fast reaction and quite similar in terms of kinetic and spectral properties to horseradish complex I.

5. Decomposition of chloroperoxidase-hydrogen peroxide complex I

The chloroperoxidase-hydrogen peroxide complex I decomposes spontaneously as shown in fig. 3, curve A. Both the formation and decomposition of complex I is measured at 410 mμ. Horseradish peroxidase complex I has been shown to decompose similarly [7]; however, in the case of horseradish peroxidase, the decomposition of complex I is accompanied by the formation of a new spectral intermediate (complex II) which absorbs at longer wavelengths (418 mμ). During the decomposition of complex I of chloroperoxidase, there is no spectral evidence for a type II complex. In fig. 3, curve B, the decomposition of complex I is measured at 430 mμ. Any complex II formation should be detectable in this region of the spectrum, however, no complex II was detected. In fact, the region between 380 and 440 mμ has been scanned and no evidence for a chloroperoxidase type II complex can be found at

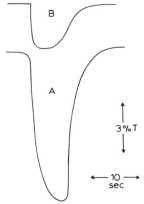

Fig. 3. *Rate of decomposition of chloroperoxidase-hydrogen peroxide complex I.* The procedures and concentrations of reactants are the same as described in fig. 2. The reaction was followed at 410 mμ (curve A) and 430 mμ (curve B) in separate experiments. In both cases, the reference wavelength was 450 mμ.

normal hydrogen peroxide concentrations. At high hydrogen peroxide concentrations (1 mM), the chloroperoxidase complex I spectrum is replaced by a new band at longer wavelengths. This new band is shifted much farther towards the red than horseradish peroxidase complex II. At this time, it is not known whether this type II chloroperoxidase complex is an active or inactive complex; however, it is clear that it is not an obligatory intermediate since it is only seen under special circumstances.

6. Dissociation constant and number of binding sites for complex I

The concentration of chloroperoxidase-complex I has been measured under varying sets of conditions with several different concentrations of hydrogen peroxide and enzyme. A Scatchard plot of this data is shown in fig. 4. The results indicate that one hydrogen peroxide molecule per enzyme molecule is required for the formation of complex I. The dissociation constant (K_d) for the formation of complex I is 2×10^{-6} M. Since the rate constant,

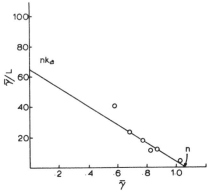

Fig. 4. *The number of binding sites and the dissociation constant for chloroperoxidase-hydrogen peroxide complex I.* The hydrogen peroxide concentration was varied between 1×10^{-5} and 5×10^{-6} M. The enzyme concentration was varied between 1×10^{-6} and 6×10^{-6} M. The titrations were carried out in a Cary 15 spectrophotometer. The maximum amount of peroxide complex was determined from the absorption loss at 410 mμ at different concentrations of peroxide extrapolated to infinite peroxide concentration. The data are plotted according to the method of Scatchard [10].

k_{-1} (see equation 2) for the back reaction (formation of free enzyme and substrate from complex I) is related to K_D and k_1 (equation 7), and values have been established for both of these parameters, k_{-1} can be calculated from the data at hand.

$$K_d = \frac{k_{-1}}{k_1}. \tag{7}$$

Using 2×10^{-6} M for the dissociation constant and 1.5×10^6 liters moles^{-1} sec^{-1} for k_1, k_{-1} calculates to be approximately 3 sec^{-1}. This value represents a maximum value for k_{-1} since it includes any contribution from the spontaneous decomposition of the peroxide complex. These results meet the criterion of irreversibility of step 1 in the reaction mechanism probably on two counts. First, it is likely that the formation of complex I involves the release of product from hydrogen peroxide. Secondly, the back reaction (k_{-1}) is so small compared to k_1 that the reaction is essentially irreversible.

7. Halide ion complexes of chloroperoxidase

Chloride, bromide, and iodide ions form spectral complexes with chloroperoxidase at pH 3 but not at pH 5. Fig. 5 shows the difference spectra for the chloride, bromide, and iodide complexes versus the free enzyme. The spectra reflect the changes observed at infinite halide concentration, where all of the enzyme is in the form of the halide complex. A summary of these spectral changes is given in table 3. Both the major troughs and peaks observed in the difference spectra increase towards longer wavelength as the ligand changes from chloride to bromide to iodide. The peak also decreases in intensity for the same series of ligands, with that of the iodide complex being visible only at very low iodide concentrations. In terms of the absolute spectra, the chloride complex shifts the Soret band toward longer wavelengths, while the bromide and iodide complexes go to shorter wavelengths.

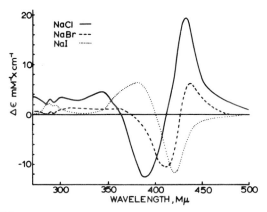

Fig. 5. *Halide difference spectra of chloroperoxidase.* The cuvette in the sample compartment of a Cary 15 recording spectrophotometer contained chloroperoxidase dissolved in 0.1 M phosphate buffer, pH 2.8 and halide ion. The reference cuvette contained chloroperoxidase and buffer only. The maximum halide concentrations used were 0.1 M for NaBr and NaCl, and 0.02 M KI. The differential extinction coefficients (E) were obtained by extrapolation of the observed changes at various halide concentrations to infinite halide concentration. At infinite halide concentration, all the enzyme is in the form of the halide complex.

Table 3
Effects of sodium halides on the spectral properties of chloroperoxidase
and horseradish peroxidase

Difference spectra			Soret maximum of the halide complex, mμ (extinction x M^{-1} cm^{-1})
Ligands	Trough (mμ)	Peak (mμ)	
H$_2$O vs H$_2$O (chloroperoxidase)	–	–	400 (75,000)
I$^-$ vs H$_2$O (chloroperoxidase)	421	442	394 (77,000)
Br$^-$ vs H$_2$O (chloroperoxidase)	412	437	394 (69,000)
Cl$^-$ vs H$_2$O (chloroperoxidase)	390	433	415 (70,000)
Cl$^-$ vs H$_2$O (horseradish peroxidase)	395– 400	413	405 (70,000) 370 (80,000)

Horseradish peroxidase has been investigated to see if it forms halide complexes at pH 3 similar to chloroperoxidase. These conditions appear to denature horseradish peroxidase, changing the visible spectrum to one resembling hemin at pH 3.

The dissociation constants for the chloroperoxidase-halide complexes were determined by making double reciprocal plots of the halide concentrations versus the changes in absorbance observed at the major troughs in the difference spectra (fig. 6). The halide ion concentrations necessary to achieve one half maximal absorbance change are 0.2 M chloride, 0.4 M bromide, and 0.015 M iodide. Although these are very weak complexes, the rate of complex formation is fast as indicated by studies using the stopped-flow technique. Preliminary studies indicate the rate constant for chloride binding to the enzyme to be approximately 1×10^4 M^{-1} sec^{-1}.

In addition to the spectral changes observed in the Soret region due to halide ion binding, spectral differences were also detected in the 280-295 mμ region. A peak at 292 mμ, trough at 288 mμ, and another peak at 284 mμ is seen in the difference spectra. Changes similar to this have been characterized as being due to perturbation of a tryptophan residue [10]. In the chloroperoxidase-

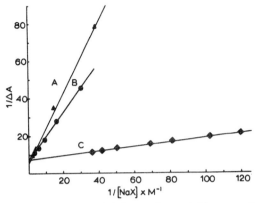

Fig. 6. *Dissociation constants for the halide complexes of chloroperoxidase.* The difference spectra between chloroperoxidase in the presence and absence of various concentrations of halide ion were measured at an appropriate wavelength for each halide complex. The reciprocal of the absorbance change was then plotted versus the reciprocal of
the halide concentration used. The dissociation constant was determined from the halide
concentration at one half the extrapolated maximal absorbance change. For the chloride
(B) and iodide (C) complexes, the absorbance changes were measured at 392 mμ and
421 mμ respectively. The enzyme concentration in both cases was 1.2×10^{-5} M. For the
bromide complex (A), the wavelength monitored was 414 mμ, and the enzyme concentration was 2.1×10^{-5} M. All measurements were made in 0.1 M potassium phosphate
buffer, pH 2.8.

halide ion complex, the perturbation would represent the tryptophan experiencing a more hydrophilic environment upon the addition of sodium halide.

A change in the heme ligand would also be expected to cause
spectral differences in the visible region of the spectrum. Only very
slight changes were found in the visible region, with peaks in the
difference spectra at 545 mμ and 575 mμ. The changes observed
were about one half the magnitude of those observed for the tryptophan perturbation.

8. Neutral and acid forms of chloroperoxidase

As indicated in the previous section, chloroperoxidase forms complexes with halide substrates at pH 3 but not in the pH 5 to 7 range. This finding correlates with the pH optimum for the halogenating activity of chloroperoxidase. On the other hand, our studies have shown that chloroperoxidase has classical peroxidase activity (classical peroxidase activity is defined herein as the ability to catalyze the peroxidation of pyrogallol, malachite green, guaiacol, and related substrates in the absence of halide ion) over a broad pH range, pH 3 to 7. Figure 7 compares the halogenating (curve 1) and classical peroxidase (curve 2) activity of chloroperoxidase at different pH values. These data together with the studies on the pH dependence of halides complexes of chlorperoxidase indicate that chloroperoxidase can exist in two different catalytically active forms. The acid form of the enzyme possesses halo-

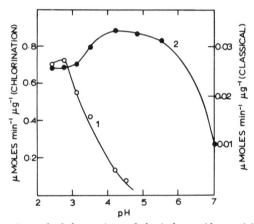

Fig. 7. *The pH optimum for halogenation and classical peroxidase activity of chloroper-* *oxidase.* Curve 1 (O———O) shows the rate of the chlorination of monochlorodimedon as measured in the standard assay conditions [1] except for the variation in pH which was accomplished by using potassium phosphate buffers with varying pH values. Curve 2 (●———●) shows the rate of guaiacol peroxidation as a function of pH. The guaiacol reaction mixtures contained 300 μmoles of potassium phosphate buffer at variable pH values between pH 2 and 7, 6 μmoles of hydrogen peroxide, 20 μmoles of guaiacol and chloroperoxidase in a total volume of 3 ml. The rate of guaiacol oxidation was determined by following the change in absorbance at 470 mμ. The rate in terms of μmoles oxidized per minute per μg of chloroperoxidase was calculated using 2.66×10^4 as the molar extinction coefficient for tetraguaiacol [11].

genating activity while the neutral form of the enzyme catalyzes classical peroxidase reactions. Definite spectral differences in the Soret region of the spectrum can be detected between the neutral and acidic forms of the enzyme (see fig. 8). The neutral minus acid difference spectrum shows a maximum at 413 mμ with a molar extinction coefficient of 2.2×10^3. In fig. 9, the magnitude of the spectral difference at 413 mμ is plotted against pH and compared to the halogenating activity (k_{cat} for the overall chlorination of monochlorodimedon) as a function of pH. The two curves show exactly parallel pH dependencies. If one assumes that the spectral difference between the neutral and acid forms of chloroperoxidase is due to the titration of a group in the vicinity of the heme group, a pK of 3.5 to 4.0 can be estimated for this group on the basis of the titrations shown in fig. 9. The most likely candidate for this group would be an aspartic or glutamic carboxyl; however, the possibility of an imidazole nitrogen of histidine having an abnormally

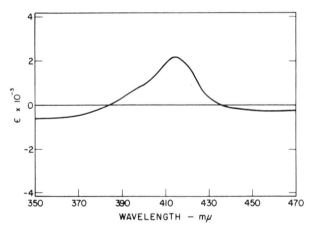

Fig. 8. *The pH difference spectrum of the acid minus the neutral form of chloroperoxidase.* The cuvette in the sample compartment of a Cary 14 recording spectrophotometer contained chloroperoxidase dissolved in 0.1 M potassium citrate-phosphate buffer, pH 5.4. The reference compartment contained chloroperoxidase dissolved in 0.1 M potassium citrate-phosphate buffer, pH 2.8. The extinction coefficients are based on the concentration of hemin in the enzyme sample.

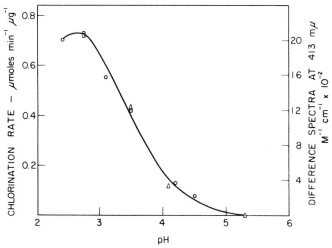

Fig. 9. *The pH dependence of the spectral difference between the acid and neutral forms of chloroperoxidase.* The spectral difference at 413 mμ (△———△) between the acid and neutral forms of chloroperoxidase was measured with enzyme dissolved in 0.1 M potassium citrate-phosphate buffer, pH 5.4 (sample compartment) and with enzyme dissolved in 0.1 M citrate-phosphate buffer at the indicated pH (reference compartment). The pH dependence of the chlorination of monochlorodimedon (○———○) was measured in the standard assay mixture [1] with 0.1 M citrate-phosphate buffer replacing the regular phosphate buffer.

low pK cannot be ignored. Westheimer [12] has preliminary information suggesting that the pK of the lysine residue at the active center of aceto-acetate decarboxylase is 4 to 5 pK units lower than that of a normal lysine. By analogy, a histidine residue in chloroperoxidase could readily fall into the necessary 3.5 to 4.0 pK range.

The interconversion of the acid and neutral forms of chloroperoxidase is very rapid. It must occur faster than 5 milliseconds as judged by the fact that it is too fast to be observed in the stop flow apparatus. No evidence for a change in the conformation or in the state of aggregation of chloroperoxidase in shifting between the acid and neutral forms could be obtained from fluorescence emission spectra or from sedimentation velocity data.

9. Chloride complex formation as a function of pH

When the extent of formation of chloride complex with the acid form of the enzyme (as measured by the difference spectrum at 430 mμ) is compared with k_{cat} for the overall halogenation reaction as a function of pH the curves parallel each other in the pH range pH 5.4 to 2.8 (see fig. 10).

The pH dependence of three processes, namely the k_{cat} for the overall halogenation reaction, the spectral differences between the acid and neutral forms of the enzyme, and the halide complexation suggest that these processes are interrelated. A suitable model to explain this relationship is readily provided by the hypothesis that halide ion and an ionizable group on the enzyme (having a pK of -3.6) compete for one of the six coordinate positions of the heme iron. This hypothesis, outlined in fig. 11, postulates that at neutral

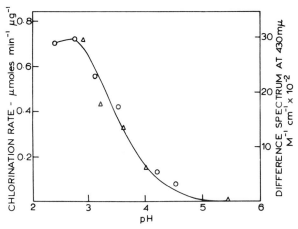

Fig. 10. *The extent of chloride complex formation as a function of pH.* The spectral difference at 430 mμ ($\triangle$———$\triangle$) due to the formation of enzyme-chloride complex was measured with chloroperoxidase dissolved in 0.1 M citrate-phosphate buffer at the indicated pH value in the presence of 0.02 M potassium chloride (sample compartment) and with chloroperoxidase dissolved in 0.1 M citrate-phosphate buffer at the same pH in the absence of chloride (reference compartment). The pH dependence of the chlorination of monochlorodimedon was measured as described in fig. 9.

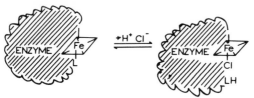

Fig. 11. *Postulated ligand change in the formation of the chloroperoxidase-chloride complex.*

pH values, an amino acid side chain of the protein donates a ligand (L^-) to the 6th coordinate position of the heme iron. However, at acid pH, this ligand becomes protonated (LH) and chloride ion now occupies the 6th coordinate position. If this hypothesis is correct, chloride ion and L^- can be considered to be competitive inhibitors of one another. Thus the chloride ion concentration which gives one half maximal velocity (K_m) in the chlorination reaction should vary as a function of the hydrogen ion concentra-

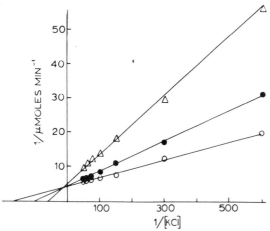

Fig. 12. *The effect of pH on V_{max} and K_m for chloride in the halogenation reaction.* The initial rate of monochlorodimedon chlorination was measured under standard assay conditions [1] using 0.09 µg of chloroperoxidase and the indicated concentrations of chloride. Citrate-phosphate buffer (0.1 M), pH 2.8 (○———○), pH 3.1 (●———●), and pH 3.5 (△———△) replaced the usual phosphate buffer in the standard assay. The data are plotted as reciprocals of rate and chloride concentration in the Lineweaver and Burk fashion.

tion, however, the maximal velocity (V_m) at infinite chloride concentration should be constant. Fig. 12 traces Lineweaver-Burk plots for chloride at pH 2.8, 3.1, and 3.5. As is evident in this plot, V_m is essentially constant over this pH range, whereas the K_m for chloride ion increases from 6.25×10^{-3} to 1.75×10^{-2} M in going from pH 2.8 to pH 3.5. These data are therefore consistent with the hypothesis of halide ion competing for a heme ligand position with an ionizable group on the enzyme having a pK in the 3.5–4.0 range.

10. Summary

This report has examined the spectral and kinetic properties of complexes between chloroperoxidase and two of its substrates, hydrogen peroxide and halide anions. This information has been fitted with other studies concerning the mechanism of enzymatic halogenation. The following general points have been established:

1. A model which is consistent with both steady state and transient state kinetic studies has been developed. The model features an ordered sequence of substrate addition to the enzyme with the first step being essentially irreversible.

2. In the first step, chloroperoxidase reacts with hydrogen peroxide in a pH independent, irreversible reaction to form a 1 : 1 complex or compound. This complex is spectrally similar to complex I of horseradish peroxidase. The second order rate constant for the formation of chloroperoxidase complex I is 1.6×10^6 M^{-1} sec^{-1}.

3. No complex corresponding to complex II of horseradish peroxidase can be detected at low hydrogen peroxide concentrations. At high concentrations of hydrogen peroxide (1 mM), a new complex (compound) is formed which is spectrally similar to horseradish complex II; however, this complex cannot be an obligatory intermediate in the halogenation reaction.

4. Chloroperoxidase exists in two forms, an acid (pH 3 – 4) and a neutral form (pH 5–7). The two forms can be differentiated on the basis of kinetic and spectral properties. The acid form of the enzyme catalyzes halogenation reactions and forms complexes with chloride, bromide, and iodide ions. The neutral form of the enzyme catalyzes classical peroxidation reactions but does not halogenate and does not form halide ion complexes.

5. The chloroperoxidase-halide complexes presumably involve interaction of the halide anion as a ligand to the heme group of the enzyme. Complex formation involves characteristic changes in the Soret absorption band. The rate constant for chloride binding to the enzyme is estimated to be $1 \times 10^4 M^{-1} sec^{-1}$; or approximately $1/200$ the rate of hydrogen peroxide binding.

6. The binding of halide to chloroperoxidase perturbs a tryptophan residue.

7. Halide binding is quite reversible, whereas peroxide binding is not.

8. Hydrogen peroxide binding is pH independent whereas the binding of halide ions shows the same pH dependence as k_{cat} for the overall reaction. Therefore, halide ions and hydrogen peroxide must occupy separate binding sites on the enzyme.

9. The pH dependency and kinetic studies on halide binding indicate that halide ion competes with an ionizable group on the protein which has a pK value in the 3.5–4.0 range. The most likely candidate for this ionizable group would be an aspartic or glutamic carboxyl group.

10. Based on these considerations, the proposed model for chloroperoxidase catalysis can be further enlarged to suggest heme iron ligands as binding sites for chloroperoxidase substrates. Fig. 13 A and B presents the details of this model. In 13 A the heme group in the neutral form of chloroperoxidase is depicted with one protein ligand and one water ligand. Reaction of the neutral form of chloroperoxidase with hydrogen peroxide leads to the formation of complex I using the water ligand heme site.

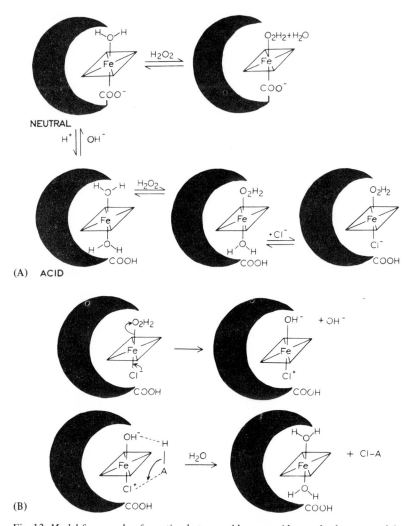

Fig. 13. *Model for complex formation between chloroperoxidase and substrates and for generation of an enzyme-bound halogenium ion.* In (A), in writing the reaction of H_2O_2 with enzyme to form complex I, we do not mean to imply that all the elements of hydrogen peroxide must remain intact in complex I. However, since the exact chemical nature of complex I is unknown, we have chosen the simplest way to express the concept of one hydrogen peroxide molecule reacting with one enzyme molecule with preservation of the 2 oxidizing equivalents of hydrogen peroxide.

The conversion of the neutral form to the acid form of chloroperoxidase is accomplished by protonation of the carboxylate heme ligand. The acid form of the enzyme can now react with both hydrogen peroxide and halides using the fifth and sixth heme iron coordinate positions as binding sites.

A two-electron oxidation of the halide ion (13 B, upper left) with the heme iron serving as a shuttle to hydrogen peroxide generates the enzyme-bound halogenium ion intermediate (13 B, upper right). Reaction of this intermediate with a halogen acceptor yields the halogenated product and regenerates free enzyme.

References

[1] D.R.Morris and L.P.Hager, J. Biol. Chem. 421 (1966) 1763.
[2] L.P.Hager, D.R.Morris, F.S.Brown and H.Eberwein, J. Biol. Chem. 241 (1966) 1769.
[3] D.R.Morris, Ph.D. Thesis, University of Illinois (1965).
[4] F.S.Brown and L.P.Hager, J. Am. Chem. Soc. 89 (1967) 719.
[5] J.A.Thomas, Ph.D. Thesis, University of Illinois (1968).
[6] W.W.Cleland, Biochim. Biophys. Acta 67 (1963) 104.
[7] B.Chance, Arch. Biochem. 41 (1950) 404.
[8] B.Chance, Arch. Biochem. 22 (1949) 224.
[9] G.Scatchard, Ann. N.Y. Acad. Sci. 51 (1949) 660.
[10] S.Yanari and F.A.Bovey, J. Biol. Chem. 235 (1960) 2818.
[11] A.C.Maehly and B.Chance, Meth. Biochem. Anal. 1 (1954) 357.
[12] F.Westheimer, personal communication.

MYELOPEROXIDASE-MEDIATED ANTIMICROBIAL SYSTEMS AND THEIR ROLE IN LEUKOCYTE FUNCTION*

S.J.KLEBANOFF

USPHS Hospital, Seattle, Washington 98114
and the Research Training Unit, Department of Medicine,
University of Washington School of Medicine, Seattle, Washington, 98105

1. Introduction

The primary function of the neutrophilic polymorphonuclear leukocyte is the phagocytosis and destruction of microorganisms. These processes have been the subject of extensive morphological and biochemical studies. Initially, cellular pseudopods surround the organism and gradually incorporate it into an intracellular vacuole lined by the invaginated cell membrane. The cytoplasm of the neutrophil contains numerous membrane-bound granules. Following phagocytosis, degranulation of the leukocyte occurs [1–6]. The membranes of the granules adjacent to the phagocytic vacuole appear to fuse with the membrane of the vacuole. An explosive rupture of the granules follows, with the discharge of their contents into the phagocytic vacuole which contains the ingested organism.

Phagocytosis and granule rupture are associated with a burst of

---

* This paper was presented at the symposium on Membrane Function and Electron Transfer to Oxygen, Miami, Florida, 22–24 January 1969.

leukocyte metabolic activity [7–12]. There is an increase in glucose uptake, lactic acid production and phospholipid turnover. Oxygen consumption is increased many times. There is a striking increase in the conversion of glucose carbon-1 to CO_2 and a lesser increase in the conversion of glucose carbon-6 to CO_2. These metabolic changes appear to result in a sharp fall in pH in the vicinity of the ingested particle [13] and in the generation of H_2O_2 by the cell [11, 14–17].

The morphological and biochemical events which follow phagocytosis normally terminate in the death of the ingested microorganisms. Acid, lysozyme which is present in the leukocyte granules [18], a variety of granular cationic proteins with antibacterial activity [19–20] and H_2O_2 [11] have been implicated. This paper will consider the role of the neutrophil peroxidase myeloperoxidase, in the antimicrobial activity of the leukocyte.

Myeloperoxidase is present in the neutrophil in exceptionally high concentrations [21,22]. It is localized in the large, azurophilic or primary granules of the neutrophil which also contain the acid hydrolytic enzymes characteristic of lysosomes [23–26]. It is released, along with the other lysosomal contents, during the degranulation which follows phagocytosis. This was first reported by Graham in 1920 [27] as follows: "Very marked changes in the granules may readily be determined by the study of leucocytes engaged in phagocytosis, as for example, in opsonic preparations or in smears of pus. Smears from an active case of gonorrhoea are very satisfactory. When such preparations are stained with a "peroxidase" reagent interesting examples of the more or less complete disappearance of the granules from individual cells may be obtained. While exceptions may occur, it may be stated in general that the granules disappear from the leucocytes progressively as the number of bacterial inclusions in the cell increases." Fig. 1 is an electron micrograph of a human neutrophil which has been stained for peroxidase following the ingestion of bacteria. Two bacteria are seen within a phagocytic vacuole. A peroxidase positive granule

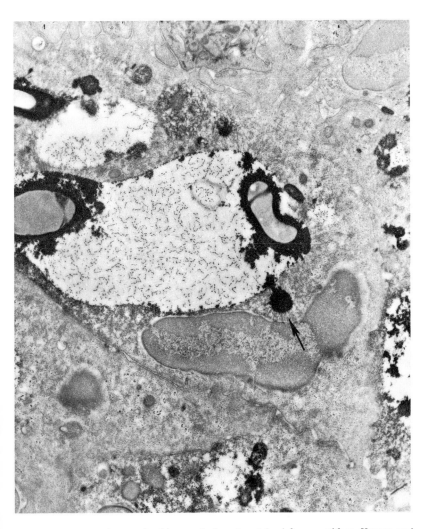

Fig. 1. Electron micrograph of human leukocytes stained for peroxidase. Human peripheral blood was incubated with bacteria (*L. acidophilus*) for 30 min. The buffy coat was fixed with 2.5% glutaraldehyde, stained for peroxidase by the diaminobenzidine method [28] and post-fixed with osmium tetroxide. The sections were stained with uranyl acetate. Magnification 10,700 X.

(arrow) appears to be discharging its contents into the vacuole, and there is a concentration of the electron dense product of the peroxidase reaction adjacent to the bacteria. Release of myeloperoxidase by the lysis of isolated granules results in an increase in peroxidase activity and in susceptibility to inactivation by H_2O_2 and aminotriazole [29] which suggests that the latency characteristic of lysosomal enzymes applies also to myeloperoxidase.

Peroxidases can exert an antimicrobial effect under certain conditions. Thus Kojima observed that peroxidase and H_2O_2 greatly increase the germicidal power of a number of phenols [30]. An antimicrobial system has been found in saliva and in milk which consists of lactoperoxidase, thiocyanate ions and H_2O_2 [31−43]. Iodide ions can replace thiocyanate ions as the heat-stable, dialyzable component of the antilactobacillus system of saliva [34, 35,37]; however, the concentration of iodide ions required is greater than that present in saliva. It is presumed that when an oxidizable cofactor such as a phenol or thiocyanate ions is required, the antimicrobial effect is due to the conversion of the oxidizable cofactor by peroxidation from a weak to a strong antimicrobial agent.

2. Myeloperoxidase-halide-H_2O_2 antimicrobial system

Klebanoff and Luebke reported in 1965 that a highly purified preparation of dog myeloperoxidase can replace lactoperoxidase in the peroxidase-thiocyanate-H_2O_2 antimicrobial system [37] which suggested that a peroxidase-mediated antimicrobial system similar to that found in saliva and milk might also be operative in the neutrophil. Thiocyanate ions can be replaced as the oxidizable cofactor of the myeloperoxidase-mediated antimicrobial system by iodide, bromide or chloride ions [44−46]. The iodinated hormones, thyroxine or triiodothyronine also can be employed; however, the deiodinated derivative, thyronine, is inactive in equivalent concentrations [45]. Thyroxine and triiodothyronine are degraded by peroxidase and H_2O_2 with the release of inorganic iodide.

Dog myeloperoxidase can be replaced by purified human myeloperoxidase. Table 1 demonstrates the bactericidal effect of human myeloperoxidase, H_2O_2 and either iodide, bromide or chloride ions on *E. coli*. The purified myeloperoxidases can be replaced by a weak acid extract (0.01 M citric acid or 0.2 M acetate buffer pH 3.5) of either intact guinea pig leukocytes [46] or isolated rabbit PMN leukocyte granules. McRipley and Sbarra [47] have reported that the antimicrobial activity of a guinea pig leukocyte granule lysate prepared by the repeated freezing and thawing of isolated granules is enhanced by the addition of H_2O_2. A halide or thiocyanate was not employed in their system. An acid extract of intact guinea pig leukocytes or purified myeloperoxidase and H_2O_2 also exert an antimicrobial effect at relatively high concentrations in the

Table 1

Bactericidal effect of human myeloperoxidase. The reaction mixture contained sodium lactate buffer pH 5.0–10 μmoles, *E. coli* – 2.0 X 10^6 organisms, water to a final volume of 0.5 ml and the supplements indicated below as follows: human myeloperoxidase-17 *o*-dianisidine units; sodium iodide – 0.005 μmole; sodium bromide – 0.05 μmole; sodium chloride – 5 μmoles and H_2O_2 – 0.005 μmole. Incubation period – 60 min at 37°C. The viable cell count was determined as previously described [45].

Supplements	*E. coli* (viable cell count)
None	2.0 X 10^6
Myeloperoxidase + iodide + H_2O_2	7.0 X 10^4
Myeloperoxidase + H_2O_2	2.0 X 10^6
Myeloperoxidase + iodide	1.8 X 10^6
Iodide + H_2O_2	1.9 X 10^6
Myeloperoxidase + bromide + H_2O_2	4.5 X 10^4
Myeloperoxidase + bromide	1.9 X 10^6
Bromide + H_2O_2	1.6 X 10^6
Myeloperoxidase + chloride + H_2O_2	3.5 X 10^4
Myeloperoxidase + chloride	1.8 X 10^6
Chloride + H_2O_2	2.0 X 10^6

absence of added halide [46]. The bacteria may be directly oxidized by myeloperoxidase and H_2O_2 under these conditions; however, the introduction of a small amount of halide or other appropriate oxidizable cofactor into the reaction mixture with the other reagents or the bacteria must be considered. H_2O_2 and iodide ions also are bactericidal at relatively high concentrations in the absence of added myeloperoxidase [45]. However, when the concentrations of the components of the complete myeloperoxidase-halide-H_2O_2 system are decreased to levels at which each alone, or in combinations of 2 has little or no effect on the viable cell count, the complete system is still strongly bactericidal ([45,46], table 1).

The H_2O_2 required as a component of the peroxidase-mediated antimicrobial system can be supplied by microbial metabolism. Thus the addition of H_2O_2 is not required for growth inhibition when a H_2O_2 generating organism such as *Strep. cremoris* [33] or *L. acidophilus* [37,46] is employed. However, when H_2O_2 or a H_2O_2 generating system such as glucose + glucose oxidase, hypoxanthine + xanthine oxidase, the autoxidation of ascorbic acid or the Mn^{++} dependent aerobic oxidation of the reduced pyridine nucleotides by peroxidase, is added in addition to peroxidase and the oxidizable cofactor, growth inhibition is extended to organisms which do not generate the H_2O_2 required for the completion of the antimicrobial system [38,45]. It is possible, in mixed cultures, for a H_2O_2 generating organism to supply the H_2O_2 required for the antimicrobial effect of the peroxidase-mediated system on a non-H_2O_2 producing organism. Table 2 demonstrates the bactericidal effect of *L. acidophilus* on *E. coli* in the presence of myeloperoxidase and iodide ions. Each component of the myeloperoxidase-iodide-lactobacillus antimicrobial system is required. The antimicrobial effect is decreased by preheating the lactobacilli at 100°C for 10 min and by the addition of catalase. The myeloperoxidase-iodide-lactobacillus antimicrobial system also is toxic to *Staph. aureus* and *S. marsescens*. Iodide ions can be replaced by bromide, chloride or thiocyanate ions and myeloperoxidase by lactoperoxidase. The

Table 2

Bactericidal effect of *L. acidophilus* on *E. coli* in the presence of myeloperoxidase and iodide ions. The reaction mixture contained sodium lactate buffer pH 5.0 – 10 μmoles; *E. coli* – 2.6 $\times 10^6$ organisms, water to a final volume of 0.5 ml and the supplements indicated below as follows: dog myeloperoxidase – 150 *o*-dianisidine units; sodium iodide – 0.5 μmole; *L. acidophilus* – 2.9 $\times 10^6$ organisms; catalase – 45 μg (1035 Worthington units).The lactobacilli and catalase were heated at 100°C for 30 min where indicated. Incubation period – 60 min at 37°C.

Supplements	*E. coli* (viable cell count)
None	2.6 $\times 10^6$
Myeloperoxidase + iodide + lactobacilli	1.9 $\times 10^4$
Myeloperoxidase + iodide	2.1 $\times 10^6$
Myeloperoxidase + lactobacilli	2.1 $\times 10^6$
Iodide + lactobacilli	2.7 $\times 10^6$
Myeloperoxidase + iodide + lactobacilli (heated)	2.3 $\times 10^6$
Myeloperoxidase + iodide + lactobacilli + catalase	2.6 $\times 10^6$
Myeloperoxidase + iodide + lactobacilli + catalase (heated)	2.5 $\times 10^4$

Table 3

The fungicidal effect of the myeloperoxidase-halide-H_2O_2 system. The reaction mixture contained sodium lactate buffer pH 5.0 – 10 μmoles, *Candida tropicalis* – 2.5 $\times 10^5$ organisms, glucose – 5 μmoles, water to a final volume of 0.5 ml and the supplements indicated below as follows: dog myeloperoxidase – 150 *o*-dianisidine units; H_2O_2 – 0.5 μmole; sodium iodide – 0.05 μmole; sodium bromide 0.5 μmole and sodium chloride – 25 μmoles. Incubation period – 60 min at 37°C.

Supplements	*Candida tropicalis* (viable cell count)
None	2.5 $\times 10^5$
Myeloperoxidase + H_2O_2 + iodide	0
Myeloperoxidase + H_2O_2	2.4 $\times 10^5$
Myeloperoxidase + iodide	2.2 $\times 10^5$
H_2O_2 + iodide	2.1 $\times 10^5$
Myeloperoxidase + H_2O_2 + bromide	9.0 $\times 10^2$
Myeloperoxidase + bromide	2.3 $\times 10^5$
H_2O_2 + bromide	2.6 $\times 10^5$
Myeloperoxidase + H_2O_2 + chloride	2.0 $\times 10^2$
Myeloperoxidase + chloride	1.9 $\times 10^5$
H_2O_2 + chloride	1.6 $\times 10^5$

antagonistic effect of one bacterial species on another through the formation and secretion of H_2O_2 has been described [48–52]. The data in table 2 indicate that this type of bacterial antagonism is greatly enhanced by the addition of peroxidase and an appropriate oxidizable cofactor.

A variety of bacterial species, both gram positive and gram negative, have been found to be susceptible to inhibition by the myeloperoxidase-mediated antimicrobial systems [38,45–47] although a systematic survey of organism susceptibility has not been done. The myeloperoxidase-mediated antimicrobial system also has fungicidal activity. Table 3 demonstrates the fungicidal effect of the myeloperoxidase-halide-H_2O_2 system on *Candida tropicalis* in lactate buffer pH 5.0. The complete myeloperoxidase-halide-H_2O_2 system was required for optimal fungicidal effect under the conditions employed. Iodide ion was the most effective halide, followed by bromide and then chloride ion, as was observed for the bactericidal effect of the myeloperoxidase-halide-H_2O_2 system [46].

The antimicrobial effect of the myeloperoxidase-halide-H_2O_2 system has an acid pH optimum (pH 4.5–5.0) and is inhibited by cyanide, azide, thiocyanate, 1-methyl-2-mercaptoimidazole (Tapazole), thiourea, reduced glutathione, cysteine, ergothioneine, thiosulfate, NADH, NADPH and catalase [45,46]. The inhibition by thiocyanate is of interest since thiocyanate also can be employed as a component of a myeloperoxidase-mediated antimicrobial system [37,38].

Myeloperoxidase can catalyze the formation of iodine-carbon covalent bonds when incubated with iodide ions, H_2O_2 and a suitable iodine acceptor molecule such as tyrosine or tyrosine residues of protein [53–56]. Bacteria appear to be an appropriate iodine acceptor substance, for when bacteria are incubated with myeloperoxidase, iodide and H_2O_2, iodide is converted to a TCA precipitable form and the iodide can be seen radioautographically to be fixed to the bacteria [45]. A number of myeloperoxidase-catalyzed reactions including the iodination reaction are stimulated by chloride ions [57,58,56]. Table 4 demonstrates the stimulation by bromide or chloride ions of the iodination of *L. acidophilus* (live or heat-killed) and albumen by myeloperoxidase and H_2O_2. Little iodination is observed in the absence of bromide or chloride ions under the conditions employed (low iodide to H_2O_2 ratio). The properties of the iodination reaction catalyzed by myeloperoxidase and those of the myeloperoxidase-iodide-H_2O_2 antimicrobial system were comparable in all the parameters measured which suggests that the death of the organisms is associated with, and may therefore be a consequence of, their iodination [45].

It is of interest that Agner had suggested that myeloperoxidase may serve a detoxification function based on the finding that crude diphtheria toxin was inactivated by myeloperoxidase and H_2O_2 [59]. Subsequent studies with purified diphtheria toxin revealed a requirement for an oxidizable, dialyzable cofactor [60], and among many which were found to be effective was iodide [53]. When iodide labelled with I^{131} was employed, the iodination of the toxin molecule was demonstrated [53].

The effect of catalase on the iodination reaction catalyzed by myeloperoxidase and on the myeloperoxidase-mediated antimicro-

Table 4

Stimulation of iodination by bromide and chloride ions. The reaction mixture contained sodium lactate buffer pH 4.5–10 μmoles, sodium iodide – I^{131} – 0.001 μmole (0.2 μc), dog myeloperoxidase – 17 o-dianisidine units, H_2O_2 – 0.1 μmole and water to a final volume of 0.5 ml. Sodium bromide – 1 μmole, sodium chloride – 100 μmoles, *L. acidophilus* (liver or heat-killed) 2×10^8 organisms and crystalline bovine albumen – 500 μg were added as indicated. Incubation period 60 min at 37°C. Iodination was determined as previously described [45].

Supplements	Iodide incorporation (percent)		
	Iodide	Iodide + bromide	Iodide + chloride
None	1.1	0.1	1.4
L. acidophilus (live)	1.0	18.3	11.5
L. acidophilus (heat-killed)	3.1	52.4	40.0
Albumen	1.5	72.5	60.0

Table 5

Iodination catalyzed by catalase. The reaction mixture contained sodium lactate buffer pH 4.5–30 μmoles, sodium iodide – I^{131} – 0.1 μmole (0.2 μc), water to a final volume of 0.5 ml and the supplements indicated below as follows: catalase 445 μg (10,325 Worthington units); glucose – 5 μmoles; glucose oxidase – 0.1 μg (0.1 Sigma unit). Incubation period – 60 min at 37°C.

Supplements	Iodide incorporation (mμmoles)
None	0.1
Catalase + glucose + glucose oxidase	59.8
Catalase + glucose	0.1
Catalase + glucose oxidase	0.1
Glucose + glucose oxidase	0.1

bial systems is of special interest since catalase may be present in leukocytes and can be formed by a number of microbial species. Catalase has an inhibitory effect on both iodination and microbial killing by myeloperoxidase when reagent H_2O_2 is employed; however when a H_2O_2 generating system is used, catalase can replace peroxidase as the catalyst of the reaction under certain conditions. Table 5 demonstrates the stimulatory effect of catalase on the conversion of iodide to a TCA precipitable form in a reaction mixture which also contains lactate buffer pH 4.5, glucose and glucose oxidase and indicates the requirement for each component of the reaction mixture. In table 5, catalase is the predominant protein and thus appears to serve both as the catalyst of the iodination reaction and as the iodine acceptor molecule. When the catalase concentration is decreased, the iodination of albumen, thyroglobulin, bacteria or tyrosine by catalase can be demonstrated. Table 6 demonstrates the bactericidal effect of catalase, iodide, glucose and glucose oxidase on *L. acidophilus* under these conditions. These studies suggest that iodide ions are among the substances which can be oxidized by catalase and H_2O_2 under appropriate conditions, i.e., at acid pH and in the presence of low steady-state concentrations of H_2O_2. When an iodine acceptor group is available, iodination can

Table 6

Bactericidal effect of catalase on *L. acidophilus.* The reaction mixture contained sodium lactate buffer pH 4.5 – 30 μmoles, *L. acidophilus* – 2.5 $\times$ 10^6 organisms, water to a final volume of 0.5 ml and the supplements indicated below as follows: catalase – 4 μg (103 Worthington units); sodium iodide – 0.005 μmole; glucose – 5 μmoles; glucose oxidase – 0.001 μg (0.001 Sigma unit). Catalase and glucose oxidase were heated at 100°C for 10 min where indicated.

Supplements	*L. acidophilus* (viable cell count)
None	2.5 $\times$ 10^6
Catalase + iodide + glucose + glucose oxidase	1.5 $\times$ 10^3
Catalase + iodide + glucose	2.5 $\times$ 10^6
Catalase + iodide + glucose oxidase	2.2 $\times$ 10^6
Catalase + glucose + glucose oxidase	2.5 $\times$ 10^6
Iodide + glucose + glucose oxidase	2.9 $\times$ 10^6
Catalase (heated) + iodide + glucose + glucose oxidase	1.8 $\times$ 10^6
Catalase + iodide + glucose + glucose oxidase (heated)	2.5 $\times$ 10^6

occur. The demonstration of a bactericidal effect of catalase under these conditions emphasizes the complexity of the relationship of catalase to the myeloperoxidase-mediated antimicrobial systems. Under most experimental conditions, i.e., at pH levels above 5.0 or in the presence of relatively high concentrations of H_2O_2, catalase would be expected to inhibit H_2O_2-dependent antimicrobial systems. However, the conditions of pH and H_2O_2 concentration present in the leukocyte following phagocytosis may closely approximate the optimum conditions for the catalase-mediated antimicrobial system and thus catalase under these conditions may contribute to the antimicrobial effect.

3. Myeloperoxidase-mediated antimicrobial systems in the intact leukocyte

When intact human neutrophils are incubated with bacteria and labelled iodide, the iodide is converted to a form which is TCA precipitable and water insoluble and the fixed iodine can be localized radioautographically in the cytoplasm of the cell in association with the ingested bacteria (fig. 2; see also [45]). No fixation of iodine is observed in the absence of intracellular bacteria or in the presence of 1 mM Tapazole [45], an agent which inhibits thyroidal iodination. Iodination by intact neutrophils also is inhibited by

1 mM azide or cyanide (fig. 2). When viewed with the electron microscope, the silver grains are most commonly seen near the surface of the ingested micro-organisms (fig. 3). However, silver grains also are seen over peroxidase-positive granules and at the periphery of phagocytic vacuoles at some distance from the intravacuolar bacteria. Iodination by neutrophils also is observed radioautographically when the bacteria are replaced by a suspension of endotoxin (Bacto-Lipopolysaccharide *E. coli* 026:B6 Boivin; *E. coli* 026:B6 Westphal;

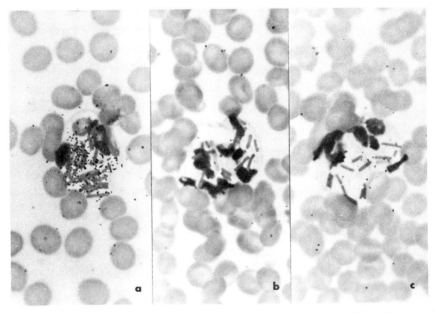

Fig. 2. Inhibition of iodination in intact neutrophils by azide and cyanide. Normal human heparinized blood was incubated with heat-killed *L. acidophilus* and Na I[125]. Sodium azide and sodium cyanide were added to b and c respectively at a final concentration of 0.001 M. There were no further additions to a. The incubation and preparation of radioautograms were performed as previously described [45]. Magnification 2250 X.

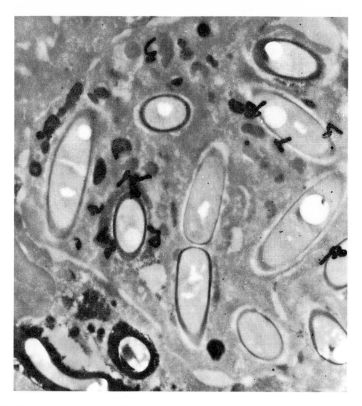

Fig. 3. Electron microscopic radioautogram of a human neutrophil following incubation with *L. acidophilus* and Na I^{125}. The section, prepared as described in fig. 1, was coated with Ilford L-4 emulsion. The section was unstained except for the peroxidase stain. Magnification 11,500 X.

E. coli 055:B5 Westphal; *S. typhosa* 0901 Westphal and *S. enteritidis* Westphal) at a final concentration of 20 µg per ml.

The iodination reaction which occurs in the intact leukocyte is believed to be largely a consequence of the reaction of myeloperoxidase, released into the phagocytic vacuole as a result of granule rupture, with H_2O_2, formed during the leukocyte respiratory burst (or by microbial metabolism when a live H_2O_2 generating organism is employed) and the added radioactive iodide. If this interpretation is correct, it indicates that the components of the myeloperoxidase-mediated antimicrobial system are capable of reaction in the intact leukocyte. Do they contribute to the antimicrobial activity of the leukocyte and if so, to what extent and under what conditions? Although a definitive answer to these questions must await further study, there is some evidence which is at least compatible with the operation of a myeloperoxidase-mediated antimicrobial system in the intact cell.

3.1. *Effect of myeloperoxidase deficiency*

Lehrer and Cline [61] have studied a patient with systemic candidiasis, an absence of detectable peroxidase in his neutrophils and monocytes and an absence of leukocyte candidacidal activity. Their studies are compatible with the involvement of myeloperoxidase in the antifungal activity of leukocytes while suggesting that there are other antimicrobial systems in leukocytes which can function in the apparent absence of myeloperoxidase to kill microorganisms. Their patient was free of severe bacterial infections. Other patients with an apparent inherited absence of myeloperoxidase were not noted to have an increased susceptibility to either bacterial or fungal infections [62,63].

3.2. *Effect of variations in H_2O_2 formation*

A characteristic property of PMN leukocytes is the formation of H_2O_2 following phagocytosis [11, 14–17]. H_2O_2 is a component of the myeloperoxidase-mediated antimicrobial systems; thus an association between altered H_2O_2 production and antimicrobial activity of leukocytes is compatible with the operation of myeloperoxidase-mediated antimicrobial systems in the intact cell. However, it should be emphasized that H_2O_2 has an antimicrobial effect at relatively high concentrations in the absence of myeloperoxidase and that this non-enzymatic antimicrobial effect is potentiated by the addition of iodide ions [45]. Thus it cannot be assumed that H_2O_2 can exert an antimicrobial effect in the leukocyte only through its association with myeloperoxidase.

The importance of H_2O_2 in the antimicrobial activity of the leukocyte, first suggested by Iyer et al. [11], has been emphasized by
Sbarra and his coworkers. They have reported that anaerobiasis decreases the bactericidal activity of leukocytes and have suggested
that this may be due to a decrease in H_2O_2 generation by the cells
[64,65]. The candidacidal activity of leukocytes also is inhibited
by anaerobiasis [61]. Mukherjee and Sbarra have reported that if
phagocytosis of E. coli by guinea pig leukocytes occurs concurrent-
ly with low doxe X-irradiation in vitro, an increase in bactericidal
activity is observed [66]. The increased killing was partially inhibited by catalase, cysteine and reduced glutathione, which led
them to suggest that X-irradiation may increase the bactericidal
capacity of leukocytes in part, through H_2O_2 production. Leukocytes isolated from X-irradiated guinea pigs were reported to have
a decreased bactericidal activity and a decrease in H_2O_2 production
following particle ingestion [67].

An impairment of H_2O_2 production following particle ingestion
is a characteristic of the leukocytes of patients with chronic granulomatous disease of childhood [68,69]. Patients with this condition
have repeated severe infections and their leukocytes have an impaired ability to kill certain microbial species [70]. Other microorganisms, e.g. certain streptococcal strains [71] or L. acidophilus
[72], are readily killed by chronic granulomatous disease leukocytes. The neutrophils of patients with chronic granulomatous disease do not iodinate normally following the ingestion of microorganisms [72]. No iodination is observed when either heat-killed
or live S. marsescens or heat-killed L. acidophilus are employed as
the ingested particle, under conditions in which normal leukocytes
exhibit extensive iodination. When live lactobacilli are employed,
iodination is observed, although it is less than normal. The blood of
the mother of the patient with chronic granulomatous disease contained 2 populations of neutrophils; one population which iodinated
well and one population which did not iodinate following the ingestion of heat-killed L. acidophilus [72]. This would be expected

from a consideration of the Lyon hypothesis for X-linked diseases.

Lactobacilli, like the streptococci and pneumococci, have been classified by Orla-Jensen as lactic acid bacteria. Lactic acid bacteria have certain characteristics in common. Of particular pertinence to this discussion is the absence of hemoproteins and the involvement of flavoproteins in terminal oxidations with the formation of H_2O_2. Although exceptions occur, in general, lactic acid bacteria are H_2O_2 generating organisms. Thus the iodination observed in chronic granulomatous disease leukocytes when live lactobacilli are employed and the ability of these leukocytes to kill lactobacilli (and streptococci) may be due to the intraleukocytic formation of H_2O_2 by the bacteria which can compensate for the defect in H_2O_2 production by the leukocyte. Although it is probable that H_2O_2 of either leukocytic or microbial origin is required as a component of an antimicrobial system in the leukocyte, it cannot be assumed that the role of H_2O_2 in the leukocyte is limited to this action. A defect in degranulation by leukocytes of patients with chronic granulomatous disease following particle ingestion has been suggested [70] although other investigators have not confirmed this finding [73, 74]. The role of H_2O_2 in the degranulation process, if any, is unknown.

3.3. Effect of inhibitors

It might be expected that agents which inhibit the peroxidase reaction in the intact cell also would inhibit the antimicrobial activity of leukocytes under conditions in which a myeloperoxidase-catalyzed reaction is required for optimum killing. An inhibition of the candidacidal activity of leukocytes by cyanide has been reported [61]. Fig. 4 demonstrates the effect of azide, cyanide and 1-methyl-2 mercaptoimidazole (Tapazole) on the killing of L. acidophilus by intact leukocytes suspended in homologous plasma. The inhibitors are present at concentrations (1 mM) which do not inhibit phagocytosis but do effectively inhibit iodination by the intact cells. Azide and cyanide have a highly significant inhibitory effect on

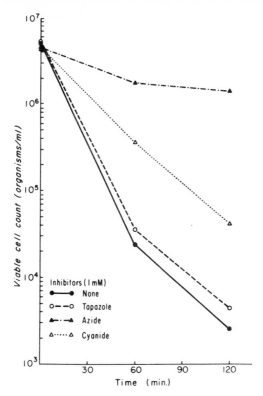

Fig. 4. Effect of azide, cyanide and 1-methyl-2-mercaptoimidazole (Tapazole) on the bactericidal capacity of human leukocytes in homologous plasma. Normal human heparinized blood was incubated with *L. acidophilus* in the absence (●——●) and in the presence of sodium azide (▲–.–▲), sodium cyanide (△. . . .△) or Tapazole (○- - -○) at final concentrations of 0.001 M and the viable cell count was determined.

bacterial killing under the experimental conditions employed whereas Tapazole has little or no effect. However, Tapazole does significantly inhibit bacterial killing under the same experimental conditions when blood from patients with chronic granulomatous disease or from carrier females is employed [72]. A danger in the use of inhibitors on intact cells is the probability that they have a more wide-spread effect on leukocyte function than simply the inhibition of peroxidase activity. The preliminary results described above need to be extended to other inhibitors and a variety of experimental conditions before they can be adequately interpreted.

4. Conclusions

What conclusions can be drawn from the studies reported here? There are probably a number of antimicrobial systems which operate in the intact leukocyte. Certain organisms may be killed by acid alone. Other organisms are sensitive to lysozyme. Heat-stable granular cationic antimicrobial factors have been described. H_2O_2, which is formed by the leukocyte, has antimicrobial properties which are potentiated by the addition of iodide ions. The antimicrobial effect of the H_2O_2-iodide system is further potentiated by the addition of myeloperoxidase. These and other studies suggest that among the antimicrobial systems in the neutrophil is one which consists of myeloperoxidase, H_2O_2 generated either by the leukocytes or by the ingested microorganisms and an appropriate oxidizable cofactor. Of the oxidizable cofactors which have been found to be effective, iodide and chloride ions are the ones most likely to be of importance in the intact cell. The uptake of iodide by normal human leukocytes has been reported to be 1200 times the uptake by erythrocytes [75] which suggests that leukocytes are among the tissues which can concentrate iodide against an electrochemical gradient. Thyroxine and triiodothyronine also are concentrated by leukocytes [75] and the deiodination of the iodinated hormones by leukocyte preparations has been reported [76]. The intracellular chloride concentration of rabbit leukocytes varies with the extracellular chloride concentration over a range for the latter of 0.3 to 161 meq. per liter of water [77]. When the leukocytes are incubated in a standard salt solution containing 115 meq. of chloride per liter, which is the concentration present in rabbit serum, the intracellular chloride concentration is 77 meq. per kg of cell water. This is considerably higher than the concentration of chloride which is required for the myeloperoxidase-mediated antimicrobial system. Some other, as yet unidentified, oxidizable cofactor may be operative under certain circumstances and it is possible that myeloperoxidase and H_2O_2 can oxidize certain microorganisms directly without the involvement of an intermediate cofactor [46,47]. The acid environment within the phagocytic vacuole [13] appears to be ideal for the operation of a myeloperoxidase-mediated antimicrobial system.

The microorganisms are killed as a result of the operation of these antimicrobial systems. It is probable that, in most instances, the entire antimicrobial armamentarium of the leukocyte is brought to bear on the ingested organism although one or other of the antimicrobial systems may predominate under certain conditions. It is suggested that the leukocytes normally have a large overkill capacity due to the variety of antimicrobial systems and the level of their activity. When this reserve is diminished as in chronic granulomatous disease leukocytes or in peroxidase-negative leukocytes, the antimicrobial effect may become more dependent upon the level of activity of one or other of the antimicrobial systems and thus more sensitive to inhibitors and activators of this system. Possibly adaptive changes can occur in leukocytes which are defective in one or other of the antimicrobial systems which can compensate for the defect. Certainly patients with a genetic absence of myeloperoxidase are less severely ill than patients with chronic granulomatous disease.

A peroxidase-mediated antimicrobial system in the intact leukocyte may be influenced by a number of factors. A decrease or complete absence of peroxidase in some or all of the neutrophils has been described in some patients with severe infections [27,78]. This deficiency is reversible if the patient survives and is thus distinguishable from the peroxidase deficiency of genetic origin. H_2O_2 production by the intact leukocyte may be influenced by the availability of oxygen [64,65] and substrate and the importance of bacteria as an ancillary source of H_2O_2 has been emphasized ([72]; table 2).

In some locations, e.g. saliva or areas of inflammation in which the leukocytes are disrupted, bacteria may be the chief source of H_2O_2 for the operation of peroxidase-mediated antimicrobial systems. If an oxidizable cofactor is required, the factors which influence the accumulation of this cofactor within the intact leukocyte also may influence the bactericidal capacity of the cell. The stimulatory effect of bromide and chloride ions on the iodination of bacteria by myeloperoxidase and H_2O_2 suggests that the halides may act synergistically to kill bacteria.

The peroxidase-mediated antimicrobial systems are inhibited by a variety of substances which occur naturally in leukocytes and bacteria. Any substance which can compete for the available H_2O_2 or can react with, and neutralize, the oxidized product of the peroxidase-catalyzed reaction may be inhibitory. The role of catalase as an inhibitor of the peroxidase-mediated antimicrobial systems requires further study in view of the finding that catalase can mimic myeloperoxidase as the catalyst of the antimicrobial system under certain conditions. The effects of exogenous agents, added to the cell suspension or administered to the intact animal are superimposed on the complexities of the naturally-occurring antimicrobial systems.

Acknowledgements

This work was supported by PHS Grant AI 07763 from the NIAID. The author gratefully acknowledges the excellent technical assistance of Mrs. Ann Waltersdorph and Miss Joanne Fluvog. Tapazole was kindly supplied by Eli Lilly and Co., Indianapolis, Ind., and the author is grateful to Dr. Julius Schultz for the purified human myeloperoxidase.

References

[1] J.Robineaux and J.Frederic, J. Compt. Rend. Soc. Biol. 149 (1955) 486.
[2] J.G.Hirsch and Z.A.Cohn, J. Exptl. Med. 112 (1960) 1005.
[3] J.G.Hirsch, J. Exptl. Med. 116 (1962) 827.
[4] W.R.Lockwood and F.Allison, Brit. J. Exptl. Path. 44 (1963) 593.
[5] D.Zucker-Franklin and J.G.Hirsch, J. Exptl. Med. 120 (1964) 569.
[6] R.G.Horn, S.S.Spicer and B.K.Wetzel, Am. J. Path. 45 (1964) 327.
[7] A.J.Sbarra and M.L.Karnovsky, J. Biol. Chem. 234 (1959) 1355.
[8] P.Elsbach, J. Exptl. Med. 110 (1959) 969.
[9] A.J.Sbarra and M.L.Karnovsky, J. Biol. Chem. 235 (1960) 2224.
[10] Z.A.Cohn and S.I.Morse, J. Exptl. Med. 111 (1960) 667.
[11] G.Y.N.Iyer, D.M.F.Islam and J.H.Quastel, Nature 192 (1961) 535.
[12] M.L.Karnovsky and D.F.H.Wallach, J. Biol. Chem. 236 (1961) 1895.
[13] P.Rous, J. Exptl. Med. 41 (1925) 399.
[14] M.Rechcigl and W.H.Evans, Nature 199 (1963) 1001.
[15] J.Roberts and Z.Camacho, Nature 216 (1967) 606.
[16] B.Paul and A.J.Sbarra, Biochim. Biophys. Acta 156 (1968) 168.
[17] M.Zatti, F.Rossi and P.Patriarca, Experientia 24 (1968) 669.
[18] Z.A.Cohn and J.G.Hirsch, J. Exptl. Med. 112 (1960) 983.
[19] J.G.Hirsch, J. Exptl. Med. 111 (1960) 323.
[20] H.I.Zeya and J.K.Spitznagel, J. Exptl. Med. 127 (1968) 927.
[21] K.Agner, Acta Physiol. Scand. 2 Suppl. 8 (1941) 1.
[22] J.Schultz and K.Kaminker, Arch. Biochem. Biophys. 96 (1962) 465.
[23] J.Schultz, R.Corlin, F.Oddi, K.Kaminker and W.Jones, Arch. Biochem. Biophys. 111 (1965) 73.
[24] D.F.Bainton and M.G.Farquhar, J. Cell. Biol. 39 (1968) 286.
[25] D.F.Bainton and M.G.Farquhar, J. Cell. Biol. 39 (1968) 299.
[26] M.Baggiolini, J.G.Hirsch and C.de Duve, J. Cell Biol. 40 (1969) 529.
[27] G.S.Graham, N.Y. State J. Med. 20 (1920) 46.
[28] R.C.Graham Jr. and M.J.Karnovsky, J. Histochem. Cytochem. 14 (1966) 291.
[29] W.H.Evans and M.Rechcigl Jr., Biochim. Biophys. Acta 148 (1967) 243.
[30] S.Kojima, J. Biochem. 14 (1931) 95.
[31] A.Portmann and J.E.Auclair, Lait 39 (1959) 147.
[32] J.Stadhouders and H.A.Veringa, Neth. Milk Dairy J. 16 (1962) 96.
[33] G.R.Jago and M.Morrison, Proc. Soc. Exp. Biol. Med. 111 (1962) 585.
[34] I.L.Dogon, A.C.Kerr and B.H.Amdur, Arch. Oral Biol. 7 (1962) 81.
[35] B.J.Zeldow, J. Immunol. 90 (1963) 12.
[36] B.Reiter, A.Pickering and J.D.Oram, Proc. 4th Int. Symp. Food Microbiol., Göteborg, Sweden (1964) p. 297.
[37] S.J.Klebanoff and R.G.Luebke, Proc. Soc. Exp. Biol. Med. 118 (1965) 483.
[38] S.J.Klebanoff, W.H.Clem and R.G.Luebke, Biochim. Biophys. Acta 117 (1966) 63.
[39] M.N.Michelson, J. Gen. Microbiol. 43 (1966) 31.
[40] J.D.Oram and B.Reiter, Biochem. J. 100 (1966) 373.
[41] J.D.Oram and B.Reiter, Biochem. J. 100 (1966) 382.
[42] Y.Iwamoto, M.Inoue, A.Tsunemitsu and T.Matsumura, Arch. Oral Biol. 12 (1967) 1009.

[43] R.R.Slowey, S.Eidelman and S.J.Klebanoff, J. Bacteriol. 96 (1968) 575.
[44] S.J.Klebanoff, J. Clin. Invest. 46 (1967) 1078.
[45] S.J.Klebanoff, J. Exptl. Med. 126 (1967) 1063.
[46] S.J.Klebanoff, J. Bacteriol. 95 (1968) 2131.
[47] R.J.McRipley and A.J.Sbarra, J. Bacteriol. 94 (1947) 1425.
[48] F.Hegemann, Z. f. Hyg. 131 (1950) 355.
[49] R.Thompson and A.Johnson, J. Infect. Dis. 88 (1951) 81.
[50] D.M.Wheater, A.Hirsch and A.T.R.Mattick, Nature 170 (1952) 623.
[51] T.Rosebury, D.Gale and D.F.Taylor, J. Bacteriol. 67 (1954) 135.
[52] R.S.Dahiya and M.L.Speck, J. Dairy Sc. 51 (1968) 1568.
[53] K.Agner, Rev. Trav. Chim. 74 (1955) 373.
[54] S.J.Klebanoff, C.Yip and D.Kessler, Biochim. Biophys. Acta 58 (1962) 563.
[55] J.-G.Ljunggren, Biochim. Biophys. Acta 113 (1966) 71.
[56] M.L.Coval and A.Taurog, J. Biol. Chem. 242 (1967) 5510.
[57] K.Agner, Abstr. Communs. 4th Congr. Biochem. Vienna (1958) p. 64.
[58] K.G.Paul, in: The Enzymes, vol. 8, eds. P.D.Boyer, H.Lardy and K.Myrbäck (Academic Press, New York, 1963) p. 227.
[59] K.Agner, Nature 159 (1947) 271.
[60] K.Agner, J. Exptl. Med. 92 (1950) 337.
[61] R.L.Lehrer and M.J.Cline, Clin. Res. 16 (1968) 331.
[62] V.J.Grignaschi, A.M.Sperperato, M.J.Etcheverry and A.J.L.Macario, Rev. Asoc. Med. Arg. 77 (1963) 218.
[63] E.Undritz, Blut 14 (1966) 129.
[64] R.J.Selvaraj and A.J.Sbarra, Nature 211 (1966) 1272.
[65] R.J.McRipley and A.J.Sbarra, J. Bacteriol. 94 (1967) 1417.
[66] A.K.Mukherjee and A.J.Sbarra, J. Reticuloend. Soc. 5 (1968) 134.
[67] B.Paul, R.Strauss and A.J.Sbarra, J. Reticuloend. Soc. 5 (1968) 538.
[68] B.Holmes, A.R.Page and R.A.Good, J. Clin. Invest. 46 (1967) 1422.
[69] R.L.Baehner and M.L.Karnovsky, Science 162 (1968) 1277.
[70] P.G.Quie, J.G.White, B.Holmes and R.A.Good, J. Clin. Invest. 46 (1967) 668.
[71] E.L.Kaplan, T.Laxdal and P.G.Quie, Pediatries 41 (1968) 591.
[72] S.J.Klebanoff and L.R.White, New Engl. J. Med. 280 (1969) 460.
[73] E.Kauder, L.L.Kahle, H.Moreno and J.C.Partin, J. Clin. Invest. 47 (1968) 1753.
[74] R.L.Baehner, M.J.Karnovsky and M.L.Karnovsky, J. Clin. Invest. 48 (1969) 187.
[75] E.Siegel and B.A.Sachs, J. Clin. Endocrinol. Metab. 24 (1964) 313.
[76] G.S.Kurland, M.V.Krotkov and A.S.Freedberg, J. Clin. Endocrinol. Metab. 20 (1960) 35.
[77] D.L.Wilson and J.F.Manery, J. Cellular Comp. Physiol. 34 (1949) 493.
[78] A.Sato and S.Yoshimatsu, Amer. J. Dis. Child. 29 (1925) 301.

MYELOPEROXIDASE DEFICIENCY: A GENETIC DEFECT ASSOCIATED WITH DIMINISHED LEUKOCYTE BACTERICIDAL AND FUNGICIDAL ACTIVITY *

Martin J.CLINE

Cancer Research Institute and Department of Medicine,
University of California Medical Center,
San Francisco, California

1. Introduction

Myeloperoxidase is a lysosomal enzyme constituting between 1 and 5 per cent of the dry weight of neutrophils [1–3]. This report summarizes the studies of a patient with disseminated candidiasis whose neutrophils and monocytes lacked myeloperoxidase, although peroxidase activity was present in his eosinophils and their precursors. These studies, which have been presented in detail elsewhere [4], confirm the genetic nature of this defect, clarify its mode of transmission, and allow an appraisal of some of the suggested biochemical and functional contributions of myeloperoxidase to the intracellular economy of the neutrophil.

* Supported by USPHS Grant No. CA 07723 and CA 11067.

2. Methods

Leukocytes were separated from heparinized venous blood of the patient, his children, and hematologically normal subjects as previously described [5]. For studies of leukocyte enzymes, the cells were washed in phosphate-buffered saline and homogenized for 60 seconds in 0.01 molar acetate buffer pH 3.8. The homogenate was centrifuged at 20,000 × g for 20 minutes and the supernatant fraction was assayed for lysozyme, cathepsin, ribonuclease, and myeloperoxidase as previously described [4]. Myleoperoxidase was measured by a modification of the method described by Klebanoff [6], using orthoanisidine as substrate. Protein concentrations were determined by the method of Lowry et al. [7].

3. Results of enzyme studies

The levels of several lysosomal enzymes in the leukocytes of normal subjects, the patient, and his family were determined biochemically. The patient's leukocyte preparations had less than 5 per cent of the normal level of peroxidase activity (table 1). The peroxidase activity of leukocytes from the patient's father and two sisters was studied only by histochemical means. One of the patient's sisters also completely lacked neutrophil and monocyte peroxidase, but had

Table 1
Leukocyte enzymes in normal subjects and a patient with myeloperoxidase-deficient neutrophils and monocytes.

	Myeloperoxidase units/µg protein	Lysozyme units/µg protein	RNA'ase mg RNA/µg protein	Cathepsin mg protein/ 30 min/10^6 PMN
Patient	<0.2	1.3	1.8	2.7
Normal subjects	2.5 ± 1	1.2 ± 0.4	1.3 ± 0.3	2.8 ± 0.7

eosinophil peroxidase activity. Her peripheral blood smear was indistinguishable from the patient's. His father and other sister have histochemically demonstrable activity in their neutrophils and monocytes, although they, like the patient's children, are presumably heterozygous. These studies are compatible with an isolated deficiency of a single enzyme in our patient's leukocytes with an autosomal recessive mode of inheritance of this trait.

4. Discussion

A 49-year-old Caucasian male was admitted to the University of California Hospitals because of systemic infection with *Candida albicans*. The patient's immunoglobulin levels, ability to produce circulating antibody, and ability to manifest delayed hypersensitivity reactions were all normal, but his leukocytes were found to have diminished microbicidal activity [4]. The patient's neutrophils were morphologically normal in fixed and stained preparations and by phase contrast and electron microscopy. These cells phagocytized fungi normally, formed phagocytic vacuoles, degranulated, and increased their oxygen utilization and glucose oxidation appropriately after phagocytosis [4]. Yet, in contrast to normal leukocytes which kill 29.0 ± 7.4 per cent of ingested *Candida albicans* in an hour, the patient's neutrophils killed only 0.1 ± 0.2 per cent. Several other species of *Candida* were tested with similar results [4]. The patient's neutrophils were also assayed in a bactericidal system containing *Serratia marcescens, Staphylococcus aureus, Escherichia coli,* or *Streptococcus faecalis.* In each instance phagocytosis of bacteria was normal, but killing was delayed. Normal neutrophils generally kill 90 per cent or more of these organisms by 45 minutes. The patient's leukocytes required 120 to 180 minutes to achieve equivalent degrees of killing [8].

Normal leukocyte microbicidal activity is inhibited by anaerobiosis, suggesting a requirement for oxygen-dependent systems.

With this clue, the patient's leukocytes were examined for peroxidase activity. His neutrophils were found to lack histochemically
detectable myeloperoxidase, and by biochemical measurements had
less than 5 per cent of the normal myeloperoxidase activity. A sister
also lacked histochemically detectable myeloperoxidase, and leukocytes from each of the patient's four sons had between 30 and 50 per
cent of the normal levels of myeloperoxidase. Activity of other
leukocyte lysosomal enzymes, including ribonuclease, lysozyme,
cathepsin, and acid phosphatase, were normal in the leukocytes of
the patient and his family.

Based on these observations of defective microbicidal activity of
myeloperoxidase-deficient leukocytes, a cell-free system based on
the studies of Klebanoff [9] was constructed. It consisted of purified myeloperoxidase, halide, and either hydrogen peroxide or a
hydrogen peroxide-generating system. The complete system was
capable of killing several strains of *Candida*. Omission of any of the
major components reduced its candidacidal activity [10].

These data define a genetic defect of neutrophil myeloperoxidase
activity which is associated with impaired microbicidal activity
against certain strains of bacteria and fungi. They lend support to
the thesis that myeloperoxidase is involved in the normal defense
systems of the human neutrophil.

References

[1] K.Agner, Acta Physiol. Scand. 2 Supp. 8 (1941).
[2] J.Schultz and K.Kaminker, Arch. Biochem. Biophys. 96 (1962) 465.
[3] J.Schultz, R.Corlin, F.Oddi, K.Kaminker and W.Jones, Arch. Biochem. Biophys.
      111 (1965) 73–79.
[4] R.I.Lehrer and M.J.Cline, J. Clin. Invest., in press, August, 1969.
[5] M.J.Cline, K.L.Melmon, W.C.Davis and H.E.Williams, Brit. J. Haematol. 15 (1968)
      539.
[6] S.J.Klebanoff, Endocrinol. 76 (1965) 301.
[7] O.H.Lowry, N.J.Rosebrough, A.L.Farr and R.J.Randall, J. Biol. Chem. 193 (1951)
      265.
[8] R.I.Lehrer, J.Hanifin and M.J.Cline, Nature, in press, 1969.
[9] S.J.Klebanoff, J. Bact. 95 (1968) 2131.
[10] R.I.Lehrer, J. Bact., in press, 1969.

MYELOPEROXIDASE AND THE BIOCHEMICAL PROPERTIES OF THE 'PEROXO-LYSOSOME' OF THE NORMAL HUMAN NEUTROPHILE

J.SCHULTZ and S. John BERGER

Papanicolaou Cancer Research Institute,
Miami, Florida 33136, USA

1. Introduction

The isolation of the peroxidase granule of the polymorphonuclear cell by density gradient techniques made possible its further purification and identification as a lysosomal particle [1–3]. Inasmuch as its characteristic enzyme is myeloperoxidase which is found in no other cell of the animal body, the name 'peroxolysosome' is suggested. Recent discoveries of diseases involving either the failure of the granule to open during phagocytosis or the lack of myeloperoxidase in the granule [4] lead us to draw attention to the properties of this particulate and the manner in which its lysis may take place.

One of the recent findings in our laboratory has been the association of lactic dehydrogenase with the peroxo-lysosome [3]. The role of LDH in this case is purely speculative. However, we believe that the lactate and DPN$^+$ produced by the phagocytic process [4], can be controlled through the LDH to form DPNH which could

thus react with MPO. In the present communication spectrophotometric evidence of marked alterations in light absorption indicated
a reaction of DPNH and MPO; MPO is inactivated and DPNH is oxidized. This reaction consumes oxygen. In the test tube DPNH generated by lactate, DPN$^+$ and LDH reacted equally well as shown
spectrophotometrically by oxygen consumption. Similar results,
obtained when experiments in which the peroxo-lysosome replaced
LDH and MPO were carried out to justify the fact that these reactions can take place in the cell.

2. Experimental and results

Oxidation of DPNH, in acid and basic media, catalyzed by MPO
is shown in fig. 1. The oxidation reaches its peak in one to two

Oxygen uptake of MPO-DPNH mixture in phosphate buffer pH 7.4.

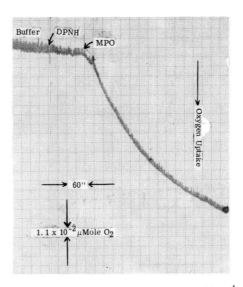

Fig. 1a. 10 mM DPNH in 0.05 M phosphate buffer pH 7.4. 3.2×10^{-1} μM MPO was
added after establishing the baseline.

Oxygen uptake of MPO-DPNH mixture in acetate buffer pH 5.2 MPO added last.

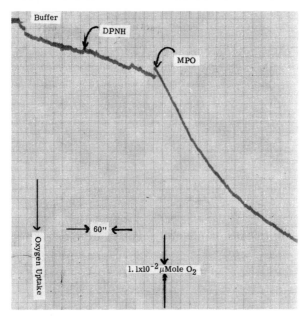

Fig. 1b. 10 mM DPNH in 0.1 M acetate buffer pH 5.2. 3.2×10^{-1} μM MPO was added to it at the arrow.

minutes at both pH's. Since DPNH is said to be unstable under acid conditions, the autooxidation of the solid and freshly prepared nucleotide was also tested at the pH's. Adding DPNH before or after MPO made no appreciable difference. The rapid initial uptake of O_2 diminished as the O_2 level in the media drops. When peroxo-lysosomes replace the MPO at equivalent MPO levels, under similar conditions as seen in fig. 2, the uptake of O_2 is even more rapid.

Inasmuch as DPNH and MPO have been shown to react, the question whether DPNH formed from lactate, LDH and DPN^+ could react with MPO and the results as shown in fig. 3 indicate active oxygen consumption. When lysosomes replaced the LDH and MPO in this system, as shown in fig. 4, oxygen consumption in these experiments at an apparently lower rate is shown in table 1, the oxy-

Oxygen uptake of MPO-DPNH mixture in phosphate buffer pH 7.4.

Oxygen uptake of MPO-DPNH mixture in phosphate buffer pH.7.4.

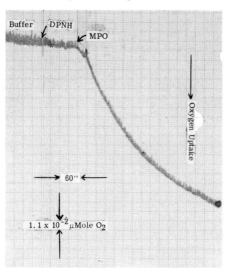

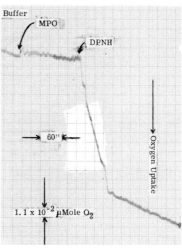

Fig. 2a. 1.8 ml 0.05 MPO$_4$ buffer pH 7.4 is used for establishing the baseline. DPNH for the final concentration of 10 mM is added. After a baseline with DPNH is established 3.2 × 10^{-1} μM MPO is added.

Fig. 2b. Same conditions and concentrations are used. DPNH is added last in this experiment.

Oxygen uptake of lysosome-DPNH mixture in phosphate buffer pH 7.4.

Oxygen uptake of lysosome-DPNH mixture in phosphate buffer pH 7.4 DPNH added last.

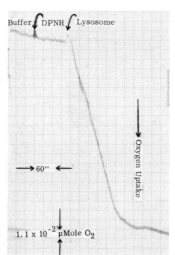

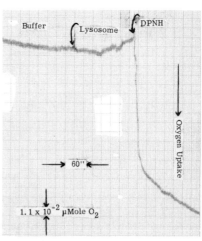

Fig. 2c. Same condition as in 2a except lysosome is used in place of MPO (concentration of MPO in lysosome is the same).

Fig. 2d. Same conditions as 2c with the exception of DPNH added last.

Oxygen uptake of MPO-DPNH mixture where DPNH is produced by lactate + LDH + DPN^+ in acetate buffer pH 5.2.

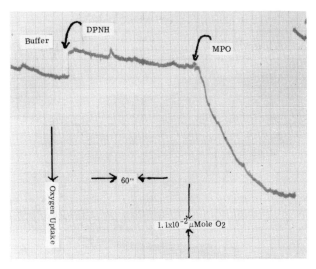

Fig. 3. 0.1 M acetate buffer pH 5.2 is used as the medium. 10 mM DPNH used in this case is produced from lactate + DPN^+ + LDH. After establishing the baseline 3.2×10^{-1} μM MPO is added.

Oxygen uptake of lysosomes in presence of lactate + DPN^+ in PO_4 buffer pH 7.4.

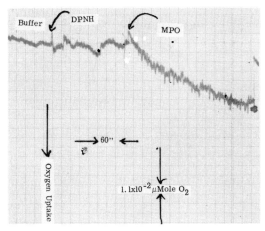

Fig. 4. 1.63 M lactate in 0.05 M PO_4 buffer is used for establishing the baseline. 10 mM DPN^+ is added. After obtaining a base line 40 λ lysosome 3.0×10^{-1} μM MPO from normal human PMN is added to it.

Table 1
Oxygen uptake of the MPO-DPNH system.

Experiment number	Lactate (M)	LDH (μM)	DPN$^+$ (mM)	DPNH (mM)	MPO (μM 10^1)	Lysosomes $\approx \mu$M MPO 10^1	Oxygen uptake μmole 10^2/5 min
1	1.65	–	–	–	–	–	0
2	1.65	4.3	–	–	–	–	0
3	1.65	4.3	6.8	–	–	–	0
4	1.65	4.3	6.8	–	2.6	–	12
5	1.65	4.3	6.8	–	1.56	–	8.8
6	1.65	4.3	6.8	–	1.3	–	6.6
7	1.65	4.3	10.0	–	1.3	–	8.3
8	–	–	–	13.6	6.5	–	24.0
9	–	–	–	13.6	2.6	–	17.6
10	–	–	–	6.8	1.3	–	7.7
11	1.65	4.3	6.8	–	–	3.25	12.0
12	1.65	–	6.8	–	–	1.5	7.7
13	–	–	–	6.8	–	0.75	5.3

gen consumption is well above the base value. These experiments suggest that the presence of LDH in the lysosomal membrane can produce DPNH for reaction with MPO in the same granule.

Maximum oxygen consumption occurs when the DPNH-MPO ratio is approximately 3.0×10^4. Below 0.7 mM DPNH is found to show no observable consumption of O_2 when $1.6 \times 10^{-1} \mu$M MPO is used. Approximately 2.0 mM DPNH is needed for $1.6 \times 10^{-1} \mu$M to obtain an observable oxidation; and the ratio is calculated to be 1.2×10^4. An increase in DPNH-MPO ratio from 3.0×10^4 to 5×10^4 does not improve the oxidation.

When denatured MPO was used in the above experiments under the same conditions using various ratios of DPNH-MPO, no oxygen consumption took place, even at a ratio where native enzyme shows the maximum oxygen uptake. DPN$^+$ alone, added either to native or denatured enzyme gave negative results.

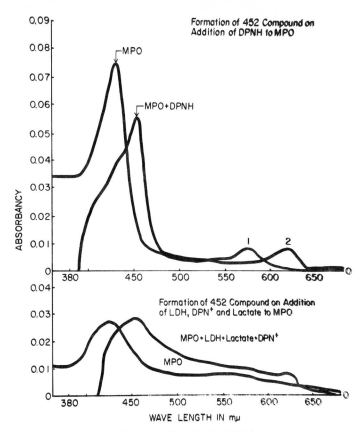

Fig. 5a. (1) 2.4 μM MPO in 0.05 M phosphate buffer pH 7.4 recorded against 0.05 M PO₄ buffer. (2) 2.4 μM MPO added to 1.34 mM DPNH in 0.05 M phosphate buffer pH 7.4 recorded against 1.34 mM DPNH in phosphate buffer. The spectrum is recorded 60 seconds after the addition of MPO to DPNH. 5b. (1) 6.1 × 10⁻¹ μM MPO in 0.05 M phosphate buffer pH 7.4 recorded against 0.05 M phosphate buffer. (2) 6.1 × 10⁻¹ μM MPO added to 0.67 mM DPNH produced by 1.6 M lactate + 8 μM LDH + 6.7 mM in phosphate buffer pH 7.4 recorded against 0.67 mM DPNH produced by the same system.

3. Spectrophotometric analysis

Oxidation of DPNH catalyzed by MPO and the spectral changes taking place in MPO were followed spectrophotometrically. Fig. 5a shows the spectra of native MPO in 0.05 M PO_4 buffer, pH 7.4, and that of DPNH-MPO complex. Fig. 5b illustrates the change taken place after the addition of DPNH produced by lactate, DPN^+ and LDH. Both spectra show the same changes. The shift of the peaks from 430 mμ and 575 mμ to 452 and 625 mμ can be noticed. The initial change occurs in the 575 region, shifting to 625 mμ. An additional peak at 375 mμ is obtained when the DPNH-MPO ratio reaches 0.406×10^3 (fig. 6). The spectral changes in both basic and acidic pH are the same. The 375 mμ peak does not appear at a ratio of 0.39×10^3. An addition of 3.0 μM of MPO to 1.34 mM DPNH, all three peaks at 625, 452 and 375 mμ are formed. But when 1.5×10^{-1} μM MPO more is added to the above mixture, both 625 and

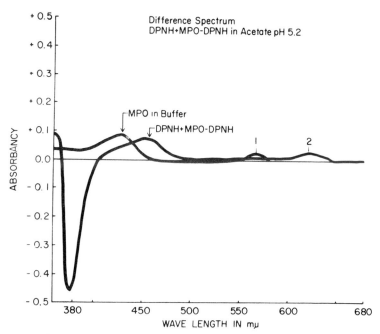

Fig. 6. (1) 1.73 μM MPO in 0.1 M acetate buffer pH 5.2 versus acetate buffer. (2) 1.73 μM MPO added to 0.95 mM DPNH in acetate buffer versus 0.95 mM DPNH in the same buffer.

452 mμ peaks have remained unchanged while an increase in the absorption is noticed at 375 mμ region. The 375 mμ peak reaches its maximum at a ratio of 0.55 $\times$ 10^3. Thus the range of DPNH concentration which causes the formation of 375 mμ peak is very narrow. 452 mμ peak fails to form when the ratio drops below 0.356 $\times$ 10^3. The initial change that takes place in the molecule is the shift at 575 mμ region to 625 mμ and once the peak is formed, it remains the same. The peak at 452 mμ starts to disappear in 90 sec to 2 min after the addition of DPNH (fig. 7). This disappearance is not followed by the reappearance of 430 mμ peak. Results obtained from the reaction carried out in 0.1 M acetate buffer pH 5.2 is shown in figs. 8 and 9 and the changes are entirely the same. Spectral changes taken place with lysosome and DPNH could not be recorded due to the mechanical difficulties.

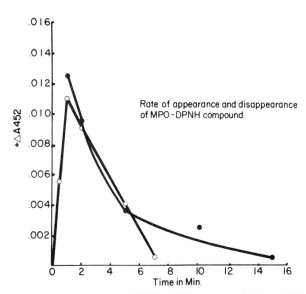

Fig. 7. Appearance and disappearance of DPNH-MPO complex with time. DPNH-MPO ratio is 0.55 $\times$ 10^3. o——o DPNH obtained from Sigma added to MPO. •——• DPNH produced from lactate + DPN^+ + LDH added to MPO.

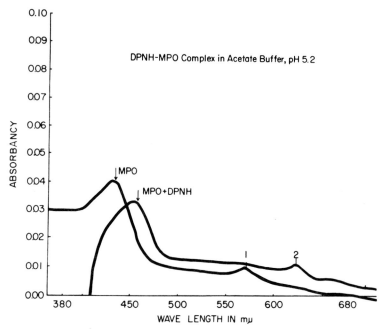

Fig. 8. (1) 8.5×10^{-1} μM MPO in 0.1 M acetate buffer pH 5.2 versus acetate buffer. (2) 8.5×10^{-1} μM MPO added to 0.47 mM DPNH in acetate buffer versus 0.47 mM DPNH in the same buffer.

The reaction of H_2O_2 and MPO is shown in fig. 10. The spectral changes seem to be the same as with DPNH-MPO complex. However, one marked difference is the reappearance of 430 mμ peak following the disappearance of 452 peak. The maximum conversion of 430 to 452 peak is shown to take place at an H_2O_2-MPO ratio of 6.1. These results show that the MPO reacts with both DPNH and H_2O_2 forming the same product. A complete conversion of H_2O_2 from DPNH may not be necessary for the action of MPO. The intermediate form HO_2- may be sufficient.

An interesting observation is the change at the 340 mμ region. Fig. 11 shows the appreciable decrease in absorbancy at 340 mμ

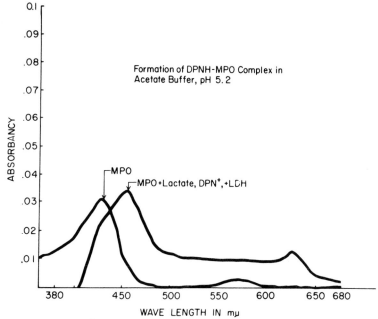

Fig. 9. (1) 8.5×10^{-1} μM MPO in 0.1 M acetate buffer pH 5.2. (2) 8.5×10^{-1} μM MPO added to 0.47 mM DPNH produced from lactate + DPN$^+$ + LDH in the same buffer.

with the addition of successful concentration of MPO. The 260 peak of DPNH remained constant.

The velocity of oxidation and the spectral change of DPN$^+$ and TPNH was also examined. DPN$^+$ does not have any effect while the TPNH-MPO spectrum showed a peak at 460 mμ. The same concentration of TPNH, as the concentration of DPNH, required for the maximum oxidation, showed no appreciable uptake of oxygen.

4. Inactivation of MPO

Various experiments were performed to study the inactivation of

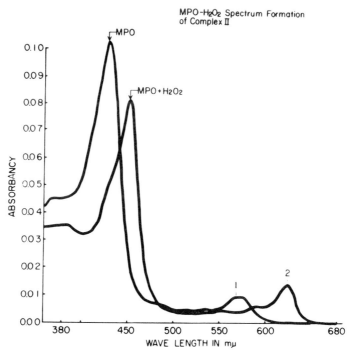

Fig. 10. (1) 2.6 μM MPO in 0.05 M phosphate buffer pH 7.4 versus 0.05 M phosphate buffer. (2) 2.6 μM MPO added to 0.44 mM H_2O_2 recorded against 0.44 mM H_2O_2 in PO_4 buffer.

MPO. The effect of H_2O_2 and DPNH (both from Sigma and from lactate, DPN^+ and LDH) on MPO were compared. The ratio required for both complexes (DPNH-MPO and H_2O_2-MPO) to inactivate at the same rate, was approximately the same. A ratio of 9×10^3 is required for 50% and 33×10^3 for 100% inactivation of MPO when both DPNH and H_2O_2 are used. Total activity is recovered when the ratio drops below 1.8×10^3. The results are duplicated when lactate, LDH and DPN^+ were used.

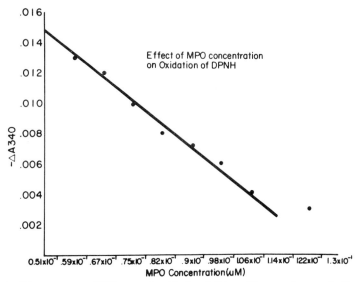

Fig. 11. Disappearance of 340 mμ peak with the addition of increased concentration of MPO. Each concentration is freshly mixed with same concentration fresh DPNH.

5. Discussion

The experiments reported here show that DPNH combines with MPO to form a complex which consumes oxygen, causes inactivation of MPO and develops spectral changes similar to those reported for other peroxidases [5]. For each experiment, the DPNH-MPO ratio is critical. No appreciable oxidation is observed at a lower concentration of DPNH whereas an excess of DPNH causes complete inactivation. The increased rate of reaction which occurred at acidic pH may provide a reason for the existence of MPO in the lysosomal fraction, although the experimental products are the same under the basic and acidic pH. The fact that the same amount of oxygen uptake was found when various lysosomal fractions obtained from a Ficoll gradient containing equal concentration of MPO explains the oxidation of DPNH is due to MPO. Lemberg and Legge [6] proposed a mechanism by which the ferroperoxidase is considered to act as oxygen binding oxidase. Yamazaki and Yokota [7,8] later divined another possible mechanism for the oxidation of DPNH by Horseradish peroxidase. According to their report, the peroxidatic reaction is produced by an original peroxidatic reaction at the expense of a trace of H_2O_2 present in DPNH solution. The presence of MPO in the lysosomal fraction leaves several unexplainable questions. The spectrophotometric data show that the DPNH-MPO complex behaves the same way as the H_2O_2-MPO complex. This observation suggests that the H_2O_2 is formed from DPNH which in turn reacts with MPO. However, the immediate formation of reaction product provides for the possible formation of HO_2- the intermediate between the oxidation of DPNH and H_2O_2 formation.

The presence of LDH in the same lysosomal fraction with MPO proposes that the lactate and the DPN^+ formed during phagocytic process of polymorphonuclear leucocytes reacts with LDH located on the granular membrane forming DPNH which on oxidation forms H_2O_2 by combining with molecular O_2, and the H_2O_2 thus formed acts as a source for the reaction of MPO. In another report in this symposium (Klebanoff) it is shown that a source of H_2O_2 is necessary for release of MPO from the granule. The experiments reported here, may or may not be part of the system, but they do show properties of the peroxo-lysosome. It may even offer a means of control of DPNH concentration in the cell.

Addendum

Since the presentation of this report, several manuscripts from this laboratory have been submitted concerning the nature of MPO. It has been found to consist of six isozymes derived from three monomeric subunits. This work originated by Felberg and Schultz, Anal. Biochem. 23 (1968) 241; Schultz, Felberg and John, Biochem. Biophys. Res. Commun. 28 (1967) 543 and Felberg, Putterman and Schultz, accepted for publication in Biochem. Biophys. Res. Commun. (1969).

References

[1] J.Schultz and K.Kaminker, Arch. Biochem. Biophys. 96 (1962) 465.
[2] J.Schultz, R.Corlin, F.Oddi, K.Kaminker and W.Jones, Arch. Biochem. Biophys. 111 (1965) 73.
[3] S.John, N.Berger, M.Bonner and J.Schultz, Nature 215 (1967) 5109.
[4] R.Baehner and M.L.Karnovsky, Science 162 (1968) 1277.
[5] T.Akazawa and E.E.Conn, J. Biol. Chem. 232 (1958) 403.
[6] R.Lemberg and J.W.Legge, Hematin Compounds and Bile Pigments, Interscience, New York (1949).
[7] K.Yokota and I.Yamazaki, Biochim. Biophys. Acta 105 (1965) 301.
[8] I.Yamazaki and K.Yokota, Biochim. Biophys. Acta 132 (1967) 310.

RESOLUTION OF SUBCELLULAR COMPONENTS OF RABBIT HETEROPHIL LEUKOCYTES INTO DISTINCT POPULATIONS BY ZONAL SEDIMENTATION AND DENSITY EQUILIBRATION

M.BAGGIOLINI, J.G.HIRSCH and C.de DUVE

The Rockefeller University, New York, N.Y. 10021, USA

1. Introduction

In a recently published paper [1] we described the resolution of subcellular components of rabbit heterophil leukocytes into three particulate fractions by zonal differential centrifugation in a B-XIV rotor [2]. Further results obtained by isopycnic centrifugation with the automatic rotor designed by Beaufay [3] as well as a summary of previously published data are presented here. The preparation of the cells, the conditions for the zonal differential centrifugation experiments, and the biochemical assays have already been described [1]. The use of Beaufay's rotor is described by Leighton et al. [3]. For the isopycnic fractionation experiments presented here, 14 ml of the starting material prepared as described previously [1] and brought to a density of 1.10 by the addition of 60% (w/w) sucrose were layered on 19 ml of a continuous sucrose density gradient extending between the densities 1.18 and 1.32 and resting on a 6 ml sucrose cushion of density 1.32. The centrifugation was then carried out at 35,000 rpm for 60 minutes. The distribution histograms of protein content and enzyme activities are presented as a function of the gradient volume for the zonal differential centrifugation and as a function of the equilibrium density for the isopycnic centrifugation experiments.

2. Results

Altogether 19 zonal differential centrifugation experiments were performed at angular velocities varying between 2,500 and 21,000 rpm. The results were very reproducible. Fig. 1 represents the distribution of four enzymes and protein in six different experiments. These four enzymes were chosen to demonstrate the four patterns of enzyme distribution found in the sedimentation experiments.

Two of these enzymes, peroxidase and alkaline phosphatase, show essentially unimodal distributions and can be separated from each other almost completely, under suitable experimental conditions (4,000 to 7,000 rpm). After centrifugation at 4,000 rpm, peroxidase is distributed in a broad, asymmetrical peak covering the outer half of the gradient and reaching the cushion, where some accumulation already occurs. Alkaline phosphatase is found in a rather narrow, symmetrical band in the inner third of the gradient, except for a tiny amount recovered at the outer limit of the gradient, and probably carried down with cytoplasmic clumps or granule aggregates as suggested by the analogous findings in experiments at 2,500 rpm. After centrifugation at 7,000 rpm, which corresponds to an approximately 3-fold increase in centrifugal force over 4,000 rpm, practically all the peroxidase is packed against the cushion, whereas the alkaline phosphatase is now recovered in the middle of the gradient. Small amounts of both enzymes, probably released from damaged particles, are found in the starting zone.

These results establish the existence of two distinct populations

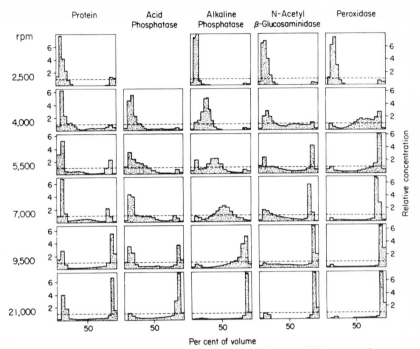

Fig. 1. Fractionation of subcellular components of rabbit heterophil leucocytes by zonal differential centrifugation at different speeds (2500–21,000 rpm) for 15 minutes. Graphs are normalized distribution histograms as a function of the volume collected. Radial distance increases from left to right. Ordinate is concentration in fraction relative to concentration corresponding to uniform distribution through the gradient. Acid phosphatase was assayed with *p*-nitrophenyl phosphate. Percentage recoveries were 97.0 ± 15.0 for protein, 78.5 ± 6.5 for acid phosphatase, 89.3 ± 9.0 for alkaline phosphatase, 91.0 ± 2.9 for β-N-acetyl glucosaminidase, 93.5 ± 4.0 for myeloperoxidase.

of particles in mature rabbit heterophil leukocytes: the A particles characterized by myeloperoxidase and the B particles characterized by alkaline phosphatase. As estimated by the median position of the two bands with respect to the starting zone, the A particles sediment, on an average, about 4 times faster than the B particles. Because of the characteristics of the sedimentation experiments, this difference is likely to be due largely to a difference in size.

β-N-acetyl glucosaminidase is distributed in two bands each carrying about 50% of the total activity, a fast sedimenting band moving with the peroxidase and a much slower sedimenting one, only slightly displaced from the starting zone at 4,000 rpm and extending broadly from the starting zone into the inner half of the gradient at 7,000 rpm. Acid phosphatase, assayed with β-glycerophosphate, has a distribution almost identical with that of β-N-acetyl glucosaminidase. The other acid hydrolases assayed (β-glucuronidase, β-galactosidase and α-mannosidase) also show this bimodal distribution. Their association with peroxidase, however, is of the order of 70–80%, the remainder being located in the broad upper band. The hydrolase activities in the upper zone of the gradient reflect for the most part the presence of particles rather than of soluble material, since centrifugation at a moderately high speed leaves only a very small proportion of the enzymes in the starting zone (fig. 1). These particles sediment much more slowly than the B particles, but considerably faster than particles of microsomal type (which would not completely accumulate against the cushion after centrifugation at 21,000 rpm) and form therefore a third distinct population (C particles). The relatively high heterogeneity in sedimentation coefficient of this population results in a fair amount of overlapping with the B particles, as well as in a poor resolution from the soluble enzymes remaining in the starting zone.

The distribution of acid phosphatase, assayed with p-nitrophenyl phosphate as substrate, is strikingly different from that of the other acid hydrolases, including the acid phosphatase determined with β-glycerophosphate. After centrifugation at 7,000 rpm, only about 10% of this acid p-nitrophenyl phosphatase is associated with the A particles, whereas most of the activity is recovered in or near the starting zone as well as in a wide band in the upper half of the gradient. Almost all of this enzyme is particle bound as shown by the distribution obtained at 9,500 and 21,000 rpm. From these results one may calculate that the sedimentation coefficient of the particles containing the acid p-nitrophenyl phosphatase is 4–5 times lower than that of the C particles. This suggests the existence of a fourth population of particles, presumably of microsomal type (D-particles), characterized by an acid phosphatase that rapidly hydrolyzes p-nitrophenyl phosphate, but has little or no activity on β-glycerophosphate.

Among the enzymes assayed, lysozyme is the only one to show a peak coinciding with that of alkaline phosphatase. However, only about two-thirds of the total lysozyme appears to be associated with the B particles. The remainder sediments with the A particles, accompanied by such small amounts of alkaline phosphatase as to make it very unlikely that an agglutination artifact is involved. Thus lysozyme is so far the only enzyme associated with both the A and the B particles.

Preliminary studies on the bactericidal activity of some of our fractions show that citric acid extracts of the A and B fractions from sedimentation experiments are both highly active in killing various bacteria *in vitro*. These extracts appear furthermore to differ significantly in their relative activities on Salmonella and Staphylococcal organisms, thus suggesting that the antibacterial agents studied by Zeya and Spitznagel [4] are distributed in these two granule types in a specific manner.

Isopycnic fractionation of the same type of starting material [1] results in resolution into four distinct particulate fractions; three of these fractions have essentially the same biochemical composition as the particle populations defined by sedimentation experiments. Fig. 2 shows the enzyme distribution histograms obtained in one of eight density equilibration experiments performed. Peroxidase and alkaline phosphatase equilibrate in two slightly overlapping bands, with modal densities 1.26 and 1.23 respectively. A small proportion of the alkaline phosphatase is regularly found at a density of 1.14–1.15, presumably reflecting the presence of enzyme bound to membrane fragments of damaged B particles.

Acid β-glycerophosphatase, β-N-acetyl glucosaminidase, β-glucuronidase and α-mannosidase are found in a sharp peak coinciding with that of peroxidase and, with the exception of α-mannosidase,

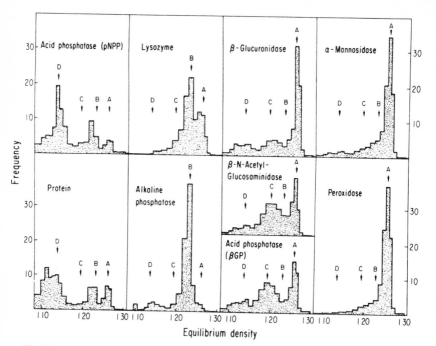

Fig. 2. Isopycnic equilibration of subcellular components of rabbit heterophil leucocytes. Graphs are frequency distributions as a function of density. Acid phosphatase was assayed with *p*-nitrophenyl phosphate (pNPP) and β-glycerophosphate (βGP). Percentage recoveries were 112 for protein, 85 for acid phosphatase (pNPP), 82 for lysozyme, 85 for alkaline phosphatase, 88 for β-glucuronidase, 91 for β-N-acetyl glucosaminidase, 120 for acid phosphatase (βGP), 91 for α-mannosidase, 92 for myeloperoxidase.

in a second, slightly broader peak with a modal density of 1.19–1.20. This second peak carries 40–50% of the total activity of the acid β-glycerophosphatase and of the β-N-acetyl glucosaminidase, but only about 20% of the β-glucuronidase, and is very likely to include the C particles defined by zonal sedimentation.

About 70% of the lysozyme equilibrates with alkaline phosphatase, whereas the remainder is found in the peroxidase rich zone. This distribution pattern confirms the dual localization of lysozyme in A and B particles.

The density distribution of acid *p*-nitrophenyl phosphatase shows a predominant peak with a modal density of 1.14, accounting for 60–80% of the total activity and accompanied by only small proportions of other enzymes. The remaining acid p-nitrophenyl phosphatase activity (20–40% of total) is distributed variably in two bands, one adjacent to the peak fraction of alkaline phosphatase, the other coinciding with peroxidase. These two bands may reflect variable agglutination of the membranous components of the D fraction to the A and B particles. In preliminary experiments, we failed to demonstrate the presence of significant amounts of 5′-nucleotidase or glucose 6-phosphatase in these particles.

In both fractionation systems, the enzyme activities of the A and B particles are associated with protein peaks. The maximum purification achieved for peroxidase is 5–7 fold in both the 7,000 rpm sediment and the fraction banding at an equilibrium density of 1.26. A purification of 4–6 fold is found for alkaline phosphatase in the B peak of density distributions as well as of zonal sedimentations carried out at 6,000–7,000 rpm.

As clearly illustrated by the density distributions of fig. 2, the A and B particles seem to contain roughly the same amount of protein. However, the protein of the B peak includes not only B particles but also overlapping A and C particles and possibly D particles carrying the acid p-nitrophenyl phosphatase. Very little protein is found in the C fraction, where β-N-acetyl glucosaminidase and acid β-glycerophosphatase exhibit high relative specific activities. The protein content of the C particles may be considerably smaller than the measured protein value, since this fraction is heterogeneous. Finally little can be said about the protein content of the D fraction since this region of the gradient is heavily contaminated by soluble cytoplasmic proteins which remain in the starting zone.

Fractions A–D from the isopycnic separation were examined in the electron microscope. Both the A and B fractions are very homogeneous, containing mostly large particles surrounded by a membrane. Both fractions also contain a fair amount of glycogen granule clusters and occasionally smooth membranes. The most striking difference between the two populations is one in size, the diameter of sharp circular profiles varying between 0.5 and 0.8 μm for the A particles, and between 0.3 and 0.5 μm for the B particles. In addition, the B particles appear to stain less intensely and to suffer manipulative damage more easily than do the A particles. The morphology of these particles is essentially the same as that described previously in A and B fractions from zonal sedimentation experiments at 6,500 rpm. The morphological properties as well as the sedimentation data identify the A particles with the azurophil, or primary granules, and the B particles with the specific, or secondary granules of rabbit heterophil leukocytes.

The C fraction is much more heterogeneous than the A and B fractions. It contains small round and irregularly shaped dense granules, a few larger granules resembling those seen in fraction B, some damaged mitochondria, smooth membranes, and glycogen granules. The numerous small dense granules seen in this fraction could represent the tertiary granules described by Wetzel et al. [5,6] and shown by these authors to stain positively for acid β-glycerophosphatase.

The D fraction contains exclusively smooth membranes in the form of morphologically empty vesicles of various size. No straight membrane fragments and practically no open vesicles are found.

3. Discussion

It has long been known that heterophil leukocytes contain two main groups of granules which differ from each other in their size and staining properties [5,7–10], and also in the time and mode of their formation during the maturation of the cells [5,10]. In the present investigation, relatively good separation of these two groups has been achieved by zonal sedimentation and isopycnic equilibration, thus allowing accurate biochemical determination of their components. In addition, two other populations of particles were identified biochemically.

According to our results, the azurophil or primary granules contain substantial amounts of five acid hydrolases regularly found in lysosomes of other cells, a sizable amount of lysozyme, very large quantities of peroxidase, and bactericidal cationic proteins.

The association of these components in one fraction after both zonal sedimentation and isopycnic equilibration, and the morphological homogeneity of this fraction argue against the possibility that two types of granules may be present in it. The recent cytochemical results of Bainton and Farquhar [11,12] showing that azurophil granules stain positively for peroxidase and five acid hydrolases are in complete agreement with our observations. The A granules may therefore be considered as a special kind of lysosomes containing, in addition to the usual hydrolases, a number of bactericidal agents like lysozyme, peroxidase [13,14] and so called cationic proteins.

Our B fraction, containing the specific or secondary granules, is not quite as pure biochemically as the A fraction with either fractionation system. Even our purest subfractions in the B zone contain trace activities of acid hydrolases. The shapes of the distributions of these enzymes, however, suggest that the specific granules, which are by far the main components of the B fraction seen by electron microscopy, are devoid of acid hydrolases, and therefore do not qualify as lysosomes. We conclude that specific or secondary granules contain essentially all of the alkaline phosphatase, as also indicated by cytochemistry [6,12], about two-thirds of the lysozyme activity of the whole cell, and bactericidal cationic proteins.

The distribution of acid hydrolases indicates that heterophil leukocytes contain a third population of particles (C particles), which qualify biochemically as lysosomes. They differ from the azurophil granules by the virtual absence of peroxidase and by their relative content of various hydrolases. Although this fraction is morphologically highly heterogeneous, it is tempting to attribute the acid hydrolase activities to the small irregularly shaped granule-like bodies with single membrane and homogeneous matrix, which are abundant in the C fraction. These granules resemble the acid phosphatase (β-glycerophosphate) positive particles described by Wetzel et al. as the tertiary granules [5,6].

By isopycnic centrifugation it was possible to resolve almost completely from the other fractions a fourth band (D), carrying most of the acid p-nitrophenyl phosphatase, and consisting exclusively of smooth vesicles. The origin of the membranous vesicles of the D fraction remains obscure; they may be derived from Golgi material, cytoplasmic vacuoles, smooth endoplasmic reticulum, granule membranes or plasma membranes. Consequently, no positive statement can be made on the localization of the acid p-nitrophenyl phosphatase in the intact cell.

Several other workers have attempted to subfractionate leukocyte granules. The design of these experiments has not permitted unequivocal conclusions, but the results obtained in them are essentially in agreement with our own findings. Working with horse blood leukocytes fractionated by conventional differential centrifugation, Vercauteren [15] has found that peroxidase, alkaline phosphatase and acid phosphatase tend, in this order, to sediment at increasing centrifugal forces. This is consistent with our results on the rabbit leukocytes, since this author used phenyl phosphate as a substrate for acid phosphatase. In his studies on horse blood leukocytes, Ohta [16] has found, again in agreement with our results, that β-glucuronidase, and also acid ribonuclease and acid protease, sediment more rapidly than do alkaline phosphatase and lysozyme. Acid phosphatase, measured with p-nitrophenyl phosphate, showed an intermediary distribution, in contrast with Vercauteren's data [15] and with our own. Schultz and coworkers have combined differential centrifugation with density gradient centrifugation in sucrose [17] or Ficoll [18] gradients to subfractionate granule fractions from human white blood cells. The results of these studies are similar to those presented here in that peroxidase was largely concentrated in a fraction of high density, together with considerable amounts of acid phosphatase and β-glucuronidase; relatively high acid hydrolase activity occurred also in a fraction of lower density in which several mitochondrial enzymes were concentrated.

In our opinion, three important technical factors contributed to the satisfactory resolution achieved in the present experiments: (a) starting cell suspensions were in every instance composed of 98% or more heterophil leukocytes, thus eliminating the possibility that one or more of the particle populations recovered might be derived from other cell types; (b) care was taken in homogenization and centrifugation procedures so that little granule breakage and essentially no granule clumping occurred; (c) zonal sedimentation, minimally complicated by artifacts, allowed full advantage to be taken of the differences in size and density between the granule populations present in the starting preparations.

Acknowledgements

This research was supported by Grant GB 5796 X from the National Science Foundation and Grant AI-01831 from the National Institutes of Health. The B-XIV rotor was put at our disposal by Dr. N.G.Anderson, Oak Ridge National Laboratory, under subcontract No. 3081, under W-7405-Eng-26, between Union Carbide Corporation and the Rockefeller University. Dr. Baggiolini was the recipient of a postdoctoral fellowship from the Stiftung für Biologisch-Medizinische Stipendien (Switzerland). The authors are indebted to Miss Annette Arcario and Mr. Armando Pelaschier for valuable technical assistance.

References

[1] M.Baggiolini, J.G.Hirsch and C.de Duve, J. Cell Biol. 40 (1969) 529.
[2] N.G.Anderson, D.A.Waters, W.D.Fisher, G.B.Cline, C.E.Nunley, L.H.Elrod and C.T.Rankin Jr., Anal. Biochem. 21 (1967) 235.

[3] F.Leighton, B.Poole, H.Beaufay, P.Baudhuin, J.W.Coffey, S.Fowler and C.de Duve, J. Cell Biol. 37 (1968) 482.
[4] H.I.Zeya and J.K.Spitznagel, J. Exptl. Med. 127 (1968) 927.
[5] B.K.Wetzel, R.G.Horn and S.S.Spicer, Lab. Invest. 16 (1967) 349.
[6] B.K.Wetzel, S.S.Spicer and R.G.Horn, J. Histochem. Cytochem. 15 (1967) 311.
[7] D.C.Paese, Blood 11 (1956) 501.
[8] Y.Watanabe, J. Electronmicr. 5 (1957) 46.
[9] M.Bessis and J.Thiery, Internat. Rev. Cytol. 12 (1961) 199.
[10] D.F.Bainton and M.G.Farquhar, J. Cell Biol. 28 (1966) 277.
[11] D.F.Bainton and M.G.Farquhar, J. Cell Biol. 39 (1968) 286.
[12] D.F.Bainton and M.G.Farquhar, J. Cell Biol. 39 (1968) 299.
[13] S.J.Klebanoff, J. Exptl. Med. 126 (1967) 1063.
[14] S.J.Klebanoff, J. Bacteriol. 95 (1968) 2131.
[15] R.E.Vercauteren, Enzymol. 27 (1964) 369.
[16] H.Ohta, Acta Haematol. Jap. 27 (1964) 555.
[17] J.Schultz, R.Corlin, F.Oddi, K.Kaminker and W.Jones, Arch. Biochem. Biophys. 111 (1956) 73.
[18] S.John, N.Berger, M.J.Bonner and J.Schultz, Nature 215 (1967) 1483.